Your Water Footprint

Your Water Footprint

The Shocking Facts About How Much Water We Use to Make Everyday Products

Stephen Leahy

FIREFLY BOOKS

A FIREFLY BOOK

Published by Firefly Books Ltd. 2014
Copyright © 2014 Firefly Books Ltd.

Introduction and conclusion Copyright © 2014 Stephen Leahy

First printing

Publisher Cataloging-in-Publication Data (U.S.)

Leahy, Stephen.
    Your water footprint : the shocking facts about how much water we use to make everyday products / Stephen Leahy.
[144] pages : col. ill., maps ; cm.
Includes bibliographical references and index.
Summary: An analysis of water usage which combines infographics with a narrative in detailing the typical volume of water necessary for common applications, from creating fuel to flushing the toilet.
ISBN-13: 978-1-77085-295-2 (pbk.)
1. Water use. 2. Water consumption. I. Title.
333.91 dc 23   TD348.L434   2014

Library and Archives Canada Cataloguing in Publication

Leahy, Stephen, 1953–, author
    Your Water Footprint : The Shocking Facts About How Much Water we Use to Make Everyday Products / Stephen Leahy.
Includes bibliographical references and index.
ISBN 978-1-77085-295-2 (pbk.)
    1. Water use. 2. Water efficiency. I. Title.
TD345.L43 2014   333.91   C2014-901161-X

Infographic research: Andrew Jones
Infographic illustration concepts: Dan Gustafson
Design: Erin Holmes
Cover design and typesetting: PageWave Graphics Inc.

Published in the United States by
Firefly Books (U.S.) Inc.
P.O. Box 1338, Ellicott Station
Buffalo, New York 14205

Published in Canada by
Firefly Books Ltd.
50 Staples Avenue, Unit 1
Richmond Hill, Ontario L4B 0A7

Printed in the United States of America

While the Publisher has made a serious attempt to obtain accurate information, the numbers represented in this book are for illustrative purposes only. Variances between sources and methodologies used for counting water footprints, area, consumption and volume may not be reflected in the text or graphics. Subsequent editions of this book may contain updated values.

The publisher gratefully acknowledges the financial support of our publishing program by the Government of Canada through the Canada Book Fund as administered by the Department of Canadian Heritage.

Contents

Introduction

Do you know you're wearing water? It takes more than 7,600 liters (2,000 gallons) of water to make a single pair of jeans and another 2,460 liters (650 gallons) to make a T-shirt. And you're eating water too. That morning cup of coffee required 140 liters (37 gallons) of water before it found its way to your table—water that was used to grow, process and ship the coffee beans. If you include toast, two eggs and some milk in your coffee, the water footprint of your breakfast totals about 700 liters (185 gallons).

Furniture, houses, cars, roads, buildings—practically everything we make uses water in the manufacturing process. When we spend money on food, clothes, cellphones or even electricity, we are buying water. A lot of water. Generating electricity from coal, oil, gas, and nuclear or hydro power involves the world's second biggest use of water after food production. Making paper is another very water-intensive process. This book required about 980 liters (260 gallons) of water to produce, or more than your morning breakfast.

We are surrounded by a hidden world of water. This unseen water is called "virtual" or "embodied" water. Even though we don't see the water it took to make a T-shirt, couch or TV, that water is just as real as the water we drink or shower with. Each of us uses far more virtual water than the "regular" water we can see, feel and taste.

According to government statistics, the average American's direct water use (the "regular" water) is around 378 liters (100 gallons) every day for showers, toilet use, washing, cooking and drinking. The virtual water in the things we eat, wear and use during a day averages 7,500 liters (1,980 gallons). That means the average American's "water footprint"—the total amount of direct plus virtual freshwater use—is about 8,000 liters (2,115 gallons) per day. Since 1 liter weighs 1 kilogram (2.2 pounds), that's the weight of four cars you have to haul if you get all that water from a well.

Due to our excessive consumption habits, a North American's daily water footprint (direct plus virtual

The Global Hydrologic Cycle

Though there is no such thing per se as "permanent removal" from the global water cycle (as all water molecules take the form of liquid, solid or gaseous water, and the cycle is unending), in this book, the term "water consumption" indicates removals that are rendered so toxic, or are relocated so far from sources of clean water, that the water is no longer accessible or fit for use.

For Every 10 People on Earth

2

have no access
to clean water
and must rely
on unimproved
sources

3

have access to an
improved water
source

5

have access to a
piped water supply
at home

Where you are in the world can dictate whether that water comes clean from a tap, from a polluted river or from a street vendor. UNICEF and the World Health Organization's Joint Monitoring Programme for Water Supply and Sanitation defines water access outside of piped water into the home as "improved" or "unimproved"; "improved" water sources include a public tap or standpipe, a tube well, a protected spring or rainwater collection. "Unimproved" water sources are unprotected wells and springs, untreated surface water, water carts or tanker trucks. Southeastern Asia and sub-Saharan Africa remain the regions with the highest numbers of people without improved sources of drinking water.

water) is more than twice the size of the global average. Think of it as running shoes placed side by side: the global shoe is a size 8, while the North American's is a size 18. By contrast, the Chinese or Indian water footprint is a tiny size 6.

THE WATER FOOTPRINT OF A BOTTLE OF COLA

To get a better understanding of virtual water and the water footprint concept, let's take a look at one popular product: a bottle of cola. Cola is almost entirely water, so a half-liter (17-ounce) bottle effectively contains 500 milliliters of water. That's the direct water input. But cola is not just water in a bottle. It contains sugar, carbon dioxide and syrup for flavoring. Sugar can be made from sugar beets, sugar cane or corn. All those crops need a lot of water to grow and to process into sugar, and the amount of water they need varies depending on where they are grown. If the sugar is made from corn grown in the United States, about 30 liters (7.9 gallons) of water was needed to grow and process the corn to make the sugar for our bottle of cola. The syrup flavoring contains small amounts of vanilla extract and caffeine from coffee beans. Vanilla and coffee require shockingly large amounts of water to grow and process. It takes about 80 liters (21 gallons) to grow and process the vanilla and 53 liters (14 gallons) to grow and process the coffee required to make just one bottle of cola.

The plastic cola bottle is made from oil. Water is crucial for the process of getting oil out of the ground and turning it into chemicals and plastics. Roughly 5 liters (1.3 gallons) of water is needed to make one half-liter (17-ounce) bottle. Then additional water is needed for packaging, shipping and so on, although these amounts are relatively small. Add all this together and the total water footprint of a bottle of cola is 175 liters (46 gallons). In other words, drinking a bottle of cola is like consuming 350 bottles of water. Stack them one on top of the other and they'd create a skinny water tower as tall as a 25-story building.

The word *consume* is used because the water footprints in this book represent amounts of water used that are not returned to an accessible location for reuse. Often water can be reused or cleaned, but the water footprint numbers here represent the net amount consumed. In other words, a water footprint is the total amount of water used, minus the clean water returned to a suitable source.

There are different ways to calculate a water footprint, so there may be differences in the numbers of liters consumed for various things. What's important is knowing that we depend on surprisingly large amounts of water in all facets of our daily lives.

Hardly anyone, including the business community and governments, is truly aware of how much water is needed to grow our food or manufacture consumer products. Yet water scarcity is already a reality in much of the world. About 1.2 billion people live in areas with chronic water scarcity, while another 2 billion are affected by shortages every year. And water scarcity is increasingly affecting people in the United States and Canada. By 2025, three in five people may be living with water shortages.

Planet Earth should really be known as Planet Water, since 70% of its surface is covered by water. About 97% of this is saltwater in the world's oceans. Of the 3% that is the planet's freshwater, 68.7% is encased in pack ice and glaciers, particularly in Antarctica and Greenland. Another 30% of our freshwater is in groundwater, and almost 1% in high-latitude permafrost.

Available freshwater is spread very unevenly across the planet. Canada has 9% of the world's freshwater, but most of that flows into the Arctic Ocean. Even with this abundance of water, 25% of Canadian municipalities have experienced water shortages. Many countries, including those in the Middle East, northern Africa, southern Europe and large parts of Asia, have relatively little water.

Water

All living things need water—humans can't survive more than three days without it. Earth has the same amount of freshwater now as it did when the dinosaurs lived here, many millions of years ago. However, two things are different today. The first difference is that most of our freshwater is frozen in the polar and Greenland ice sheets. The other difference is that we have found countless uses for water that the dinosaurs never dreamed of. In fact, water is far more useful and important in our lives than oil. At current rates of water consumption, we will experience more frequent regional water shortages as population, agricultural and industrial demands outstrip renewable supplies.

If the total amount of water on the planet hasn't changed, where does the water go when there are droughts and water shortages? Drain a lake dry to irrigate a cotton field, and that's the end of that particular water source. In other words, all the water in the lake has been consumed. Some of this lake water is in the cotton and the rest went into the atmosphere through evaporation and transpiration. Eventually that evaporated water from our drained lake will fall as rain or snow, but it will fall somewhere else, perhaps over an ocean on the other side of the planet. Over time the lake may refill, but it may take decades if the local ecosystem has been badly damaged.

The 2014 Global Water Summit concluded that shortage of water is the biggest challenge the global economy faces. It predicted that everyone on the planet will experience some serious water-related event—a shortage, a flood, an infrastructure failure, interruption to business, economic disruption—within the next 10 years.

41
Countries

abstained in voting on
U.N. Resolution 64/292,
recognizing access to clean
water and sanitation as a
basic human right

On July 28, 2010, through Resolution 64/292, the United Nations General Assembly explicitly recognized that clean drinking water and sanitation are basic human rights. The resolution called upon states and international organizations to provide financial resources and to help countries, in particular developing nations, provide safe, clean, accessible and affordable drinking water and sanitation for all. It was adopted by a recorded vote of 122 in favor, none against, and 41 abstentions. Among those abstaining were New Zealand, Israel, Denmark, Japan, the United Kingdom, the United States and Canada. Those nations that abstained reportedly determined that the declaration was premature and in the wrong forum, and that the meaning of such a right in international law was unclear.

Lakes account for just 0.26% of global freshwater, while all Earth's mighty rivers amount to only a tiny 0.006%. Together rivers and lakes represent only 1/375th of all the freshwater on the planet. That's like a parking lot with 374 red cars and a single, lonely blue one representing all the world's rivers and lakes.

A water molecule—H_2O—is made up of three atoms: two hydrogen and one oxygen. That simple configuration has near magical properties. You can freeze it, melt it, heat it and evaporate it. Almost anything can be dissolved in water. We often forget that Earth is literally a closed system, like a vessel in outer space. The total amount of water we have now is the same as it was a billion years ago. Water cannot be manufactured; it can only be moved around. We're very good at moving water around by using pipelines and canals. We're not so good at acknowledging that moving water around always means that some other place will then have less water.

Water is in constant motion. It evaporates from seas and continents, rains down from the clouds and flows from land to ocean through runoff. This is called the hydrological or water cycle. When rain falls, some of it evaporates and returns to the atmosphere; some is absorbed by soil and then taken up by plants. With enough rain, water runs off into a stream or river. Eventually all river water ends up in the world's oceans.

For a river basin or watershed to be sustainable, the amount of water we use should be no more than 20% of the precipitation that falls within it. Why not more? First, some of it evaporates. The warmer the temperatures and the drier the air, the more water evaporates. The mighty Great Lakes, containing about 20% of the world's surface freshwater, have experienced falling water levels in the past decade, mostly because of increased evaporation caused by warmer winter temperatures and little ice cover. Less than 1% of the water in the Great Lakes is renewed each year by rain and snow. The rest is 12,000-year-old water from melting of the ice sheet that once covered much of North America.

The same applies to most groundwater, including aquifers (i.e., natural underground water storage areas). We cannot take more than its natural rate of recharge from precipitation or those sources too will eventually empty, resulting in land subsidence, sinkholes or, if near coastal areas, flooding with seawater. Deep aquifers recharge at a prohibitively slow rate, and fully confined aquifers are considered non-renewable. This water is a one-time gift of nature; once emptied it cannot refill on a human time scale.

To prevent too much water from being taken, withdrawals must be limited to its "sustainable yield." A water body's sustainable yield is the amount of water that can be taken or used without having a negative impact. The second reason for limiting water withdrawals is that nature needs the rest to maintain healthy ecosystems that provide us with vital services. Nature's "green machine" of forests, wetlands and shoreline vegetation both store and clean water, not to mention cleaning the air and producing oxygen.

Humanity faces difficult choices about how best to use the limited amount of water that we have. This has become even more challenging with growing demands on water from a rising population that's expected to add a billion more people by 2030. The era of abundant, low-cost water is fading with growing water shortages, rising costs for the infrastructure that delivers water to our homes and industries, and the reality that everything we consume or use—from electricity to smartphones—needs water. Even relatively water-rich regions face water challenges, all too often because of poor water management and wasteful policies.

By knowing how dependent we are on water, not only for our health but also for our modern lifestyle, we can reduce waste, change habits and make water-smart decisions about our purchases to save water and money. Ultimately water use is all about choice. We can choose to use our water for drinking, growing food, making clothes, cars, highways, buildings, electronics, or generating energy. And we can choose to do these things using less water.

This book is not intended to list all the water footprints of everything you use in your daily life. The usefulness of the water footprint concept is that it illustrates the depth and breadth of our dependency on water in every aspect of our lives. Water footprints tell us that each North American consumes nearly 3 million liters (0.8 million gallons) of water a year. And that makes it easier to understand how changes in water availability such as drought can have local, regional, national and even globe-spanning impacts.

ENERGY

Energy is the second largest user of freshwater, totaling 15% of global water use. Water and energy are inseparable. Water is used to generate energy, and energy is used to pump and distribute water into our homes and industries and to irrigate fields. The demand to produce more energy drives up the use of water, and the demand for more water drives up energy use. According to the World Bank, by 2035 global energy consumption will have increased by 35%, while water use for energy will go up 85%.

Transportation

America's overall use of energy for transportation is astonishingly inefficient. It is estimated that 79% of the energy used for transportation is wasted, according to research by the Lawrence Livermore National Laboratory. Oil is the biggest single source of energy in the U.S.—double the amount of coal used—and almost all that oil is used just to push cars around.

Reliable data on how much water the oil industry consumes to extract, ship and process oil is not readily available. Calculating how much water pollution results is also difficult. Now that most of the easily accessible oil is gone, the industry is relying increasingly on deep-sea sources, shale oil and oil from tar sands. Deposits of oil and gas shale require a drilling technique called "fracking." This involves pumping between 7 and 23 million liters (2 to 6 million gallons) of water, along with special chemicals, down a drill hole under high pressure to crack the rock formations and release the underground oil and natural gas. Wells are often fracked several times, using millions of liters each time.

But there is more to fracking than the consumption of huge amounts of water, since much of this "frack water" will be effectively removed from the planet's water cycle. Nearly all the water used in fracking for oil or gas will become contaminated by the chemicals the industry puts in the water and by radioactive elements (such as radium) and salts found naturally in deep rock formations. Only about 5% to 10% of fracking wastewater is cleaned and reused; the other 90% to 95% is rendered unusable and pumped deep underground. Yet at present there are no regulations to restrict this practice's highly consumptive, pollution-heavy methods.

Canada is now the world's fifth largest crude oil producer and the biggest supplier of oil to the United States. Most of this oil now comes from its oil sands, also called tar sands. There is no liquid oil in the oil sands, only a tarry bitumen mixed with sandy soil deep under the pristine forests and wetlands of northern Alberta. Containing an estimated 170 billion barrels, this is the third largest crude oil reserve on the planet. As with other unconventional fossil fuels, such as shale oil and shale gas, getting the bitumen into pipelines requires extreme measures that use colossal amounts of water, energy, heat, chemicals and machinery. The oil companies—among the

richest corporations in the world—pay nothing for this water.

Like fracking, tar sands operations pollute most of the water they use and then store it in toxic wastewater lakes. These lakes now cover 176 square kilometers (68 square miles)—three times the area of New York City's Manhattan—while the companies continue to generate more than 200 million liters (53 million gallons) of this toxic liquid every day. These wastewater lakes are leaking about 11 million liters (3 million gallons) of toxic waste into rivers and the surrounding land each and every day.

This means that your car burns water—a lot of water. The World Economic Forum estimates that one liter (about 1 quart) of gasoline from conventional oil sources consumes about 3 liters (0.8 gallon) of water. Unconventional oil such as that from Canada's tar sands needs up to 55 liters (14.5 gallons) of water to produce that single liter of gasoline.

If your gas contains 10% ethanol, the water footprint of a tank of gasoline skyrockets. Ethanol consumes 1,780 liters (470 gallons) of water per liter if it's made from corn grown in the United States. If a 60-liter (16-gallon) gas tank holds 54 liters (14 gallons) of gasoline and 6 liters (1.6 gallons) of ethanol, the total water footprint of this fuel is a whopping 10,860 liters (2,869 gallons). That's enough water to fill an above-ground swimming pool 3.7 meters (12 feet) in diameter.

Biodiesel made from soybeans has an even bigger water footprint, averaging more than 8,800 liters (2,325 gallons) of water per liter for U.S. biodiesel. The world average is higher still, at more than 11,000 liters (2,906 gallons). And these numbers are only for the water it takes to grow the soy or corn, not the large amounts needed for processing. Why so much water? Green plants aren't "energy-dense," so it takes a lot of kernels of corn—2.4 kilograms (5.3 pounds)—to make a liter of ethanol. If that amount of corn were being eaten, it'd be more than enough to feed a family of four for a day.

About 45% of the U.S. corn crop is now used to make ethanol because in 2007 Congress passed a law requiring oil companies to blend billions of gallons of ethanol into their gasoline. This is likely to increase substantially in coming years unless there is a change in the law. And the United States is not alone in mandating biofuel use; 60 other countries do as well. Canada requires 5% ethanol in gasoline and 2% biodiesel in fuel and heating oil. India has a 20% biodiesel target for 2017. China plans to use 15% biofuel for all of its transportation. The European Union has a target of 7% biofuel in transport by 2020.

The International Energy Agency projects that by 2030, 5% of all road transport may be powered by biofuels. That could consume as much as 20% of all the water currently being used by global agriculture. In some countries this could be even higher, with up to 40% of the water currently used to grow food going to biofuels.

Ethanol, biodiesel and the other biofuels are still being touted as "green energy" sources and promoted by governments and others as a way of reducing emissions of carbon dioxide (CO_2) from burning fossil fuels that are causing climate change.

St. Lawrence River

To accomodate a 1 cm drop in water levels in the St. Lawrence River

6 twenty-foot-equivalent shipping containers must be taken off a container vessel

The St. Lawrence River is a prime gateway to North America, to Montreal, Toronto and the American heartland. Over 100 million tons of cargo are shipped by boat along the river every year. However, water levels fluctuate seasonally, and low levels can disrupt the flow of container traffic: a 1 cm drop in water levels means a container vessel must shed 54 tonnes in order to pass and safely berth at the Port of Montreal. This translates to a minimum of six 20-foot containers. In August 2012, water levels were more than 30 cm (1 foot) below normal for several weeks. This caused prolonged delays for importers and exporters; with profits of up to $3,500 per container for some commodities, shipping lines were hit where it hurts most.

However, as a number of scientific studies have shown, biofuel production is not a sustainable practice. One reason is that when agricultural land and water are used for fuel, there is less land and water to grow food for a hungry and thirsty world. The ethanol and biodiesel boom played a significant role in driving up food prices around the world in 2007, and it continues to do so, according to the World Bank.

Unfortunately, biofuels are making environmental problems such as climate change and water and food availability worse, not better. This is the result of setting energy policies without considering their potential impacts on water, soil, food production and natural ecosystems. This failure to understand interconnections is so widespread and the impacts so acute that institutions such as the World Bank, the World Economic Forum and the United Nations are calling on political leaders at all levels to change the way decisions are made. Truly sustainable solutions are possible only by carefully considering the linkages between clean water, growing enough food, meeting energy needs, conserving nature and reducing the emissions causing climate change. This is called the "water-energy-food nexus," and it's humanity's biggest-ever challenge.

Electricity

In the United States more water is used (but not consumed) to keep the lights on than to grow food. Plants that produce electricity, whether fueled by coal, natural gas or nuclear power, turn water into steam to drive turbines. This accounts for more than 40% of the freshwater taken from rivers, lakes and groundwater in the U.S. Most of the water is returned to the lakes and rivers, so it is not consumed and is thus not part of electricity's water footprint. (To re-emphasize, in this book we're concerned only with the amount of water consumed, because it is not returned to a suitable source and cannot be reused or repurposed.)

Most of the hundreds of millions of liters that flow through power stations every minute of every day are returned to their source. What's not so good is that it goes back as hot water. Older power plants use "once-through" cooling systems, which continuously suck in millions of liters of water per minute and spit out hot water into the lake or river where it came from. This kills lots of fish and other aquatic wildlife. Nearly all of this damage can be avoided with closed-cycle cooling systems that recirculate cooling water. Power plants built in the past 30 years use closed-cycle cooling, but the many older plants do not. However, both types of cooling systems still consume a lot of water—for a typical coal plant, 4 to 15 billion liters (1 to 4 billion gallons) of water a year.

The average American household used 10,800 kilowatt-hours (kWh) of electricity in 2012. In that year 37% of U.S. electrical generation came from coal, 30% from gas, 19% from nuclear energy, 7% from hydro power and 3.5% from wind. If your household electricity came from a coal-fired power plant, then 23,800 liters (6,300 gallons) of water were consumed to generate it. About one-third of the water was used for mining and processing the coal, and the rest evaporated at the power station.

There are many different ways of calculating energy and water consumption. The water footprints cited here of electricity from different power sources are based on research by the World Economic Forum Water Initiative. In general, energy production and distribution are high-consumption water uses. Electricity from nuclear energy takes more water than coal, because much more water is used to process uranium ore into nuclear fuel; generating 10,800 kWh takes about 35,000 liters (9,200 gallons). The latest combined-cycle natural gas power plant consumes much less, at 7,000 liters (1,800 gallons).

Surprisingly, hydroelectric power has quite a significant water-for-electricity footprint: it requires 183,000 liters (48,300 gallons) to power the average home. The reason for this is the evaporation of water in dam reservoirs. In dry regions with far higher evaporation rates a household might consume as much as 3 million liters (800,000 gallons) to meet its annual energy needs. Lake Mead, the giant Colorado River reservoir held back by the Hoover Dam, outside Las Vegas, has a surface area of 659 square kilometers (254 square miles). It loses 1.1 trillion liters (293 billion gallons) a year to evaporation. In July 2014, Lake Mead sunk to its lowest level since its creation in 1937.

Electricity generated by wind and solar power do not require water (though of course building the infrastructure does). The exception is large solar concentrators, which use mirrors to reflect the sun's heat onto a central water tower to generate steam and power a turbine. They can consume water at the rate of a nuclear station. However, some solar concentrators now use far less. The world's largest solar thermal plant, the Ivanpah project in the Mojave Desert near the border between California and Nevada, uses dry cooling. It will use the same amount of water as two holes on a golf course while producing enough power for 100,000 homes.

The United States now has the capacity to generate 61,000 megawatts of wind power, with turbines in 39 states. Another 12,000 megawatts' worth of wind power is currently under construction. The expansion of wind energy helped reduce consumption of water for electricity by 154 billion liters (36.5 billion gallons) in 2013, or about 490 liters (116 gallons) of water per U.S. resident. Equally important, wind power generation reduced CO_2 emissions from the power sector, totaling 95.6 million tonnes (105 million tons) in that year alone. This is equivalent to taking 16.9 million cars off the road, according to the American Wind Energy Association.

Choices about energy and water in an era of global warming are extremely important when nearly half the planet lives in energy poverty. More than a billion people in the developing world do not have access to electricity; 2.7 billion people rely on wood or dung for cooking and heating. When it comes to water, the truly green and sustainable sources of renewable energy are solar and wind. Even cloudy Germany generates 20% of its electricity this way. In 2013, 44% of all newly installed electrical power generation came from renewable sources—not including large hydro dams. This is a very important development; it not only reduces CO_2 emissions, it also dramatically reduces the amount of water being used to generate energy.

A number of new studies calculate that renewable energy could meet 100% of our electrical and transportation needs by 2050. In 2014 energy experts at Stanford University published a "roadmap" showing how the United States could achieve this goal. Their plan envisions a mixture of energy sources, including 55% solar, 35% wind (onshore and offshore), 5% geothermal and 4% hydroelectric. Nuclear power, ethanol and other biofuels are not included in the proposed energy mix.

The fastest, easiest and best way to reduce the water footprint of energy is to use less energy. Efficiency gains of up to 50% are feasible. Experts say the United States has more potential for improving energy efficiency than any other country. Simply by using currently available efficiency technologies, the U.S. could save more energy by 2020 than is used annually by all of Canada.

Paper versus Cloth

To produce a typical paper napkin, it takes about 13 liters (3 gallons) of water. About 60% of that water is precipitation to grow the trees and the rest is water for processing and to absorb the wastes.

Different types of trees grown in different climates naturally have varying water footprints. Eucalyptus trees consume more water than boreal pines to produce the same amount of paper. Paper is made by grinding trees into a pulp and then breaking down the wood fibers with often toxic chemicals such as chlorine. The process uses and pollutes lots of water. After that the dried pulp is sent to a paper mill for final processing.

Chlorine-free pulp processing can dramatically reduce water usage, creating "total chlorine-free bleaching" (TCF) paper products. Paper recycling can also reduce the water footprint of paper. In the United States about 30% of the total paper pulp produced is from recycled paper. In Canada it's quite low, at 20%, reflecting both the relative abundance of forests and the fact that most of the country's paper production is exported, so little of it is returned for recycling in Canada. European countries such as Germany average over 50%.

Now to tackle the age-old question: cloth or paper? Which has the smallest water footprint? A one-time-use paper napkin takes 13 liters (3 gallons) of water to make in Germany; it's probably closer to 15 liters (4 gallons) if made in the United States, with its lower recycled paper content.

A cloth napkin contains 28 grams (1 ounce) of cotton. If the cotton was grown in the United States, the water footprint of a cloth napkin would be 224 liters (59 gallons). This napkin can be washed and reused 50 times. Assume that one napkin's share of the wash water is 0.25 liter (8.5 fluid ounces), adding 12.5 liters (3.3 gallons) to the total. But since the wash water is sent for treatment, it can be reused and is not part of the water footprint.

Here's the bottom line: cloth wins easily. On a per use basis, the water footprint of paper is 15 liters (4 gallons), while cloth uses 4.5 liters (1.2 gallons). Napkins made from hemp or flax (linen) can reduce this to as low as 1 liter (0.25 gallon) per use.

AT HOME

A good water-efficient toilet uses 3.75 liters (1 gallon) per flush, compared to the 16 or 23 liters (3.5 or 5 gallons) used by older toilets. While you're in the bathroom, you might ponder where the water came from. There are only two sources: surface water (rivers and lakes) and groundwater. In urban areas, water first goes to a treatment plant and then is piped into your home. That means all water that reaches the household is treated, including your shower water, kitchen faucet water and even toilet water.

The average person flushes 80 liters (21 gallons) of high-quality drinking water down the drain every day. This formerly clean water becomes "black water" when it leaves our homes via sewers. Water from the shower, bath, dishwashing and laundry is considered "gray water"; it could be reused for some purposes without treatment, but normally it all goes into the same sewer pipe.

In urban areas, black and gray water are usually cleaned up and dumped back into a river, a lake or the ocean. Unfortunately, cities and towns in Canada and the United States regularly dump hundreds of billions of liters of raw sewage into waterways and oceans. Washington, DC, puts about 7.6 billion liters (2 billion gallons) of raw sewage into the Anacostia and Potomac Rivers each year. Like more than 700 cities in the United States, Washington uses the same pipes for both sewage and rain runoff. A heavy rainfall is often enough to overwhelm treatment systems, and the overflow is dumped, untreated, into the nearest water body. In New York City a mere 1.3 millimeters (.05 inch) of rain can overload the system.

While toilets represent the biggest direct use of water indoors—more than taking showers or washing clothes—outdoor water use for gardens, washing cars and watering lawns can account for at least half, and often far more, of our daily direct consumption of water. We could easily reduce our indoor and outdoor water use by 70% with readily available products such as water-saving toilets and showerheads, high-efficiency clothes washers and simple drip-irrigation systems.

Astonishingly, drinking and cooking account for only 1% to 2% of the roughly 375 liters (100 gallons) each of us uses directly on a daily basis. That's a mere 4 to 8 liters (1 to 2 gallons) out of the 8,000-liter (2,100-gallon) daily water footprint for the average North American. That individual daily footprint adds up to a whopping 2.9 million liters (766,100 gallons) a year, and more than 95% of it is hidden or virtual water.

FOOD

How watery do you think your breakfast is? A small (125-milliliter/4-ounce) glass of orange juice

---

443 Million

Number of school days lost each year to water-related illnesses

Which is equivalent to an entire school year for all 7-year-old children in Ethiopia

According to the United Nations Development Programme, water-related illnesses translate into the loss of 443 million school days each year. In many parts of Africa these illnesses go hand-in-hand with a lack of sanitation facilities in schools, and they are particularly acute among girls. In Uganda only 8% of schools have working toilets and only one-third have separate facilities for girls—deficits that help to explain why the country has found it difficult to reduce dropout rates among girls after puberty. Ensuring that every school has adequate water and sanitation with separate facilities for girls, making sanitation and hygiene part of the school curriculum, and equipping children with the knowledge they need to reduce health risks would go far toward reclaiming 272 million of those lost days.

needs about 200 liters (53 gallons) of water to grow, clean and process the oranges. That breaks down roughly into 150 liters (40 gallons) to grow the oranges, 40 liters (11 gallons) for processing and 10 liters (2.6 gallons) to absorb the waste. Grapefruit and apple juice require 135 liters (35.7 gallons) and 228 liters (60.2 gallons) respectively for a same-sized glass. A glass of milk takes about 200 liters (53 gallons)—the water needed to grow the food for the dairy cows and to process the milk.

This is not to say that using 200 liters of water to make a glass of OJ or milk is unsustainable. The oranges and the feed for the cows can be grown using rainfall or by withdrawing amounts of river or groundwater that are limited to the source's sustainable yield (the amount that can be taken without causing depletion).

Now consider your breakfast toast, which is probably made from wheat. In the United States, where some wheat crops are irrigated, it takes 2,200 liters (581 gallons) to grow 1 kilogram

(2 pounds) of grain. Canadian wheat is grown in a cooler climate, with little or no irrigation, so only 1,567 liters (414 gallons) of water is needed. Not surprisingly, wheat grown in the desert areas of Morocco uses more than twice as much. With a global average of 1,825 liters, the water footprint of the toast on your plate is 112 liters (30 gallons).

If you have two eggs with your toast, they required about 400 liters (106 gallons) of water, almost all of which went to grow the feed for the chickens. Add two pieces of bacon for another 300 liters (79 gallons). Finally, a cup of black coffee required 140 liters (37 gallons) to grow, process and ship the coffee beans.

Your breakfast water footprint, with a glass of either juice or milk, totals 1,152 liters (304 gallons). That's about 8 bathtubs of water for just one meal.

One way to lower your water footprint is to avoid or reduce the amount of meat you eat. Meats are far more water-intensive than grains and vegetables on a calorie-for-calorie basis. It takes 2.15 liters

Bottled Water

Every second, 1,500 bottles of water are opened in the United States. Even though it costs 1,000 to 10,000 times more than tap water, the amount of bottled water consumed each year amounts to the equivalent of every man, woman and child in America drinking 227 500-milliliter (17-ounce) bottles of water.

Plastic bottles are made from oil, and water is crucial in the process of getting oil out of the ground. Water is also critical in the process of turning oil into chemicals and plastics. Just making a half-liter plastic bottle consumes 5 liters water. Add in the 500 milliliters (17 fluid ounces) of water in the bottle, and the water footprint of a bottle of water becomes 5.5 liters (1.45 gallons). And this doesn't include the water used to generate the energy to clean and ship bottled water. When it comes to a glass of tap water, what you see in the glass is 98% of its water footprint. The only hidden water is the energy it took to move the water to the tap.

An incredible 357 billion liters (15 billion gallons) of water is consumed to make the plastic bottles Americans drink from every year. Recycling single-use plastic bottles can reduce water use, but recycling rates are still quite low, at 31% in the United States in 2012. Rates are higher in Canada—70%—because 97% of Canadians have access to curbside recycling. Not surprisingly, states with deposit programs have 48% recycling rates compared to non-deposit states, with just 20%.

In North America as much as 40% of bottled water is just tap water. Bottling companies pay a fraction of a cent per liter and then resell the water after they've run it through some filters. In blind taste tests between leading U.S. brands and New York City tap water, tap water usually wins. Nor is bottled water safer or less contaminated. In fact, tap water is more closely regulated than bottled. If there's a problem with a local source of tap water—which is tested daily—the public must be informed. The U.S. Food and Drug Administration (FDA) requires only weekly testing of bottled water; if there is a problem the company must remove the product, but it doesn't have to tell anyone. Keep in mind that FDA rules apply only to bottled water that moves from one state to another. The majority of bottled water stays in state and is regulated by local officials—or not. Ten states have no regulations at all for bottled water.

Animal Feed

37% of cereals grown in the world are used for animal feed

According to the Food and Agriculture Organization of the United Nations (FAO), some 670 million tonnes of cereals are used as livestock feed each year, or just over 37% of total world cereal use. This figure can be as high as 40% in the U.S. and as low as 15% in Africa. Depending on the cereal crop, it can exceed even those proportions: 45% of wheat grown in the E.U. is destined for animal feed. The amount of water required to grow these crops contributes to the virtual water footprint.

(0.5 gallon) of water to produce a food calorie from pork, compared to just 0.5 liter (17 fluid ounces) for cereals such as wheat and corn.

To further downsize the water footprint, replace your coffee with black tea—at 35 liters (9.2 gallons) a cup—drop the bacon and eat one egg instead of two, thus cutting out a substantial 700 liters (185 gallons) from a 1,152-liter (304-gallon) breakfast.

Meat

America's favorite lunch—hamburger and fries— weighs in with a water footprint of 2,550 liters (673 gallons). That's enough water to fill a tank as big as a full-sized SUV such as the Cadillac Escalade. It's mostly water for the beef in the burger. A medium-sized burger takes 2,350 liters (621 gallons) and medium fries add 193 liters (51 gallons). If you want to cut down on this extravagant liquid lunch, swap the beef for a soy patty and you'll subtract a whopping 2,100 liters (555 gallons).

Meat is very water-intensive because of the amount of water needed to grow the crops used to feed the animals. Beef has far and away the largest water footprint, clocking in at 15,400 liters (4,070 gallons) per kilogram (2 pounds), on average. In terms of calories, beef requires more than 10 times more water per food calorie than vegetables and cereals.

North Americans are among the world's biggest meat-eaters, averaging 83 kilograms (184 pounds) per person in 2004. This fell nearly 10% to 75.3 kilograms (166 pounds) in 2012. Higher prices and a weaker economy are thought to be the reasons for the decline. An eating revolution may also be underway as the idea of "meatless Mondays" begins to take hold, driven by health concerns. For a family of four, meatless Mondays could mean 43 kilograms (95 pounds) less meat per year, resulting in enough water savings to fill more than a dozen large tanker trucks.

If the same family decided that meatless Mondays were not for them but served chicken instead of beef throughout the year, they'd reduce their water use by an astonishing 900,000 liters (240,500 gallons), or about 30 large tanker trucks. That's a far bigger water saving than could be achieved by installing a low-flow showerhead, which could reduce home water use by 11,000 to 12,000 liters (2,900 to 3,170 gallons) per year.

An incredible one-third of the planet's land area is devoted to feeding animals we eat. Another 10% or so is used to grow food crops for humans and fiber crops such as cotton. That's more than 40% of all land area devoted to producing our food and fabrics. There isn't a great deal more land available that could be used for farming, given all the mountains, deserts, too-cold regions, urban areas and other unsuitable lands such as forests.

Water used to produce feed for food animals is often a combination of water from irrigation and from rainfall. This may or may not be sustainable, depending on how and where the animals are raised. The bottom line is that meat production requires an enormous amount of land and represents 30% of humanity's water footprint. Global meat production today is 7 times greater than in 1950. With severely limited land and water availability, many believe that we are rapidly approaching "peak meat," even though there will be another billion mouths to feed by 2030.

About one-third of the average North American diet comes from animal products (meat, fish, eggs and dairy). That translates into a daily food water footprint of 3,600 liters (951 gallons). Switching to a vegetarian diet that includes dairy products results in a much smaller water footprint of 2,300 liters (608 gallons). That's like going from a 36-inch waist to one of just 26.5 inches.

Shockingly, North Americans throw away up to 40% of all food, a 50% increase since 1970. This is an incredible waste of the water, land, labor, energy and fertilizer that it took to grow the food. A big reason why there is so much waste is confusion over product labeling. "Best before," "use by" and "best by" dates on food products have nothing to do with health or food safety. Manufacturers simply decide how long their products will remain at peak quality.

Food waste also costs each household an estimated $2,275 annually. To cut down on this, start by going easy on impulse and bulk food purchases. It's easier to curb waste by buying more often instead of purchasing massive cartloads that are hard to keep track of. Preserve food by freezing instead of leaving it in the fridge to spoil, or prepare smaller portions. When eating out, share meals—the average portion has already been supersized.

CLOTHING

Food, energy and direct water use amount to just over half of our water footprint. The rest is in the stuff we buy and fill our homes with—furniture, electronics and clothes. If you picked up the average home and dumped out its contents, it'd make quite a pile. By some estimates there'd be something like 10,000 items in it. The water footprint of all this stuff is enormous: 40,000 liters (10,600 gallons) for a lovely rug; 136,000 liters (36,000 gallons) for a leather couch; 120,000 to 200,000 liters (31,700 to 52,800 gallons) for a big flat-screen TV.

The textile and clothing industries are some of the world's biggest water consumers, standing fourth behind the food, energy and paper industries. On an average day you might wear 20,000 liters (5,300 gallons) worth of water. That's underwear (900 liters/238 gallons), socks (500 liters/ 132 gallons), a shirt (2,500 to 3,000 liters/660 to 800 gallons), jeans (8,000 liters/2,100 gallons) and a pair of leather shoes (8,000 liters/2,130 gallons). To really understand the water footprint of clothing and what this means, let's look at the life cycle of a pair of jeans.

Nearly all of the 8,000 liters (2,100 gallons) consumed by a single pair of jeans is used for growing cotton. The actual amount of water needed to grow 1 kilogram (2 pounds) of cotton varies tremendously among countries and regions. A kilogram of cotton grown in China, the world's biggest cotton producer, uses 6,000 liters (1,600 gallons), compared to 22,000 liters (5,800 gallons) in India, the second biggest producer and a major exporter of cotton. U.S. cotton has a water footprint of 8,100 liters (2,140 gallons) per kilogram, just under the global average of 10,000 liters (2,640 gallons).

Your jeans may well be made from cotton grown in the Indus Valley, in Pakistan and northeast India.

The Indus is a huge valley and river system that drains part of the Himalayan Mountains and is the birthplace of cotton. River and groundwater are used to irrigate rows of cotton plants. Your pair of jeans contains about 800 grams (28 ounces) of cotton, and thus around 15,000 liters (3,960 gallons) of water from the region. Most of the water evaporated or was used by the cotton plants, and some ended up as wastewater or gray water. That raw cotton is shipped to an urban center or to another country such as Bangladesh, the biggest exporter of textiles. In the factory the raw cotton is washed, dyed and then washed again. If you think about it, putting on our clothing is like wearing some of the water, soil and sun of faraway places such as the Indus Valley in Pakistan, and the labor of the hard-working hands in the cotton mills of Dhaka, Bangladesh.

Cotton is not only a very thirsty crop, it uses 24% of the world's insecticides, even though it represents less than 3% of the world's cropland. This harms the local ecosystem and is not healthy for farmers and their families. Although organic cotton accounts for less than 1% of all cotton produced, more and more is becoming available every year. This is cotton grown without the use of insecticides and chemical fertilizers, and thus it has a smaller water footprint because it produces little water pollution. In 2012 H&M, the large international clothing retailer, was the world's number one buyer of organic cotton. Alternative natural fibers such as hemp and flax need far fewer chemicals. They have a smaller water footprint of 2,700 and 3,780 liters (713 and 999 gallons) respectively per kilogram.

LOOKING AHEAD

The world's water use has tripled in the past 50 years. That growth rate cannot continue. Our buying choices are also water choices. Buying a product because it is "green" or uses a lot less water is the way to go, but only if it is something you truly need. The best advice is still to reduce, reuse and recycle.

Your Water Footprint will help you understand the unseen water that our modern society is built on. Keep in mind that this book is not intended to be a complete list of water footprints for everything you use in your daily life, but rather to provide an overall picture of how much water is consumed and thus no longer available for other uses.

Understanding water footprints can help explain everything from the rising cost of coffee to the price of your favorite jeans. It also goes some way toward revealing how water can be at the root of many conflicts in the world. Our hope for *Your Water Footprint* is that it will give you enough information to make water-wise choices about the things we do and what we buy. By being water-wise we can reduce our water footprint, which will save us money, help us be better prepared and more resilient during times of water shortages, and do our part to ensure that our children and grandchildren will inherit a healthy planet where fresh water is abundant.

"How inappropriate to call
this Planet Earth when it
is quite clearly Ocean."

—Arthur C. Clarke

The Big Picture

In terms of surface area, about two-thirds of our planet is covered in water. But as we'll see only a very small amount of that water is usable or readily accessible for drinking, irrigation and manufacturing. The salt contained in the oceans' water needs to be removed for it to be usable and, while it can be done through a process called "desalination," the process is expensive and requires tremendous amounts of energy. Even with seemingly vast sources of fresh water in such bodies of water like Lake Baikal in Siberia, the deepest lake in the world, and Lake Superior, the largest, the amount of water we take is far more than what is replaced naturally. Combined with elevated evaporation due to climate change, fresh water is an increasingly rare resource. Agricultural use of water taken from deep in the ground is sucking these unseen sources dry and farmland around the world is being abandoned because of water scarcity. One certainty is uncertainty: The future of our freshwater supplies is perilous.

One-third of the Earth's surface is covered by the Pacific Ocean

Pacific Ocean Facts

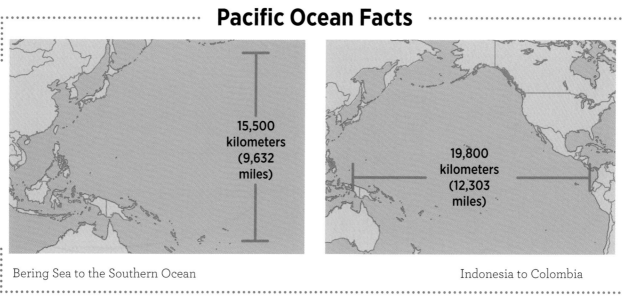

15,500 kilometers (9,632 miles)

Bering Sea to the Southern Ocean

19,800 kilometers (12,303 miles)

Indonesia to Colombia

The Pacific Ocean is the world's largest body of water. It covers 46% of the Earth's water surface, two Atlantic Oceans would fit into it, and it is larger than all the landmasses on Earth combined. Its length extends approximately 15,500 kilometers (9,600 miles), from the Bering Sea to the northern extent of the Southern Ocean, and its width is approximately 19,800 kilometers (12,300 miles), from Indonesia to the coast of Colombia—nearly halfway around the world.

If the world's water filled an **18-liter (5-gallon)** water-cooler bottle

·········· *The available freshwater would be equivalent to* ··········

Only 3 teaspoons of drinkable water

97.5% of all the water on Earth is saltwater. The remaining 2.5% is freshwater, but because almost all of it is locked up in polar icecaps, glaciers, snow and permafrost, only a tiny fraction of that is available.

Put another way, if the world's water filled an 18-liter (5-gallon) water-cooler bottle, the available freshwater would contribute only three teaspoons.

The deepest lake in the world

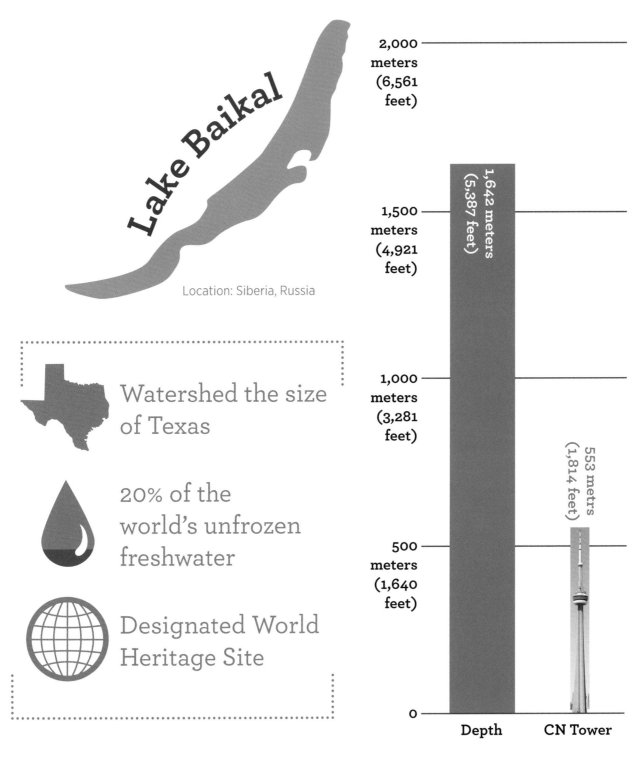

Lake Baikal

Location: Siberia, Russia

Watershed the size of Texas

20% of the world's unfrozen freshwater

Designated World Heritage Site

2,000 meters (6,561 feet)

1,500 meters (4,921 feet)

1,000 meters (3,281 feet)

500 meters (1,640 feet)

0

1,642 meters (5,387 feet)

553 metrs (1,814 feet)

Depth   CN Tower

Lake Baikal in Siberia, at 1,642 meters (5,387 feet) deep, is the deepest lake in the world. It is deep enough to hold Canada's CN Tower nearly three times over. Fed by more than 336 rivers, the crescent-shaped lake has a watershed the size of Texas. Unchecked pollution from industry and warming of the lake led to the U.N.'s declaring it a World Heritage Site in 1996, to preserve its ecological character.

The largest lake in the world

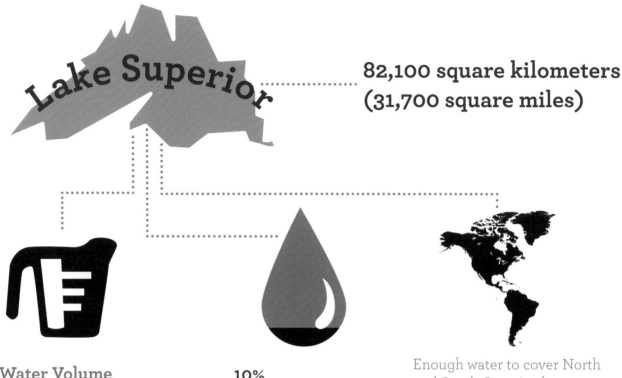

Lake Superior

82,100 square kilometers (31,700 square miles)

Water Volume
12,100 cubic kilometers (2,903 cubic miles)

10%
of the world's surface freshwater

Enough water to cover North and South America by
30 centimeters (12 inches)

If you added the surface area of

Vermont + New Hampshire + Massachusetts + Connecticut + Rhode Island

they would not equal the surface area of Lake Superior

Young in geological terms (less than 10,000 years old), Lake Superior is not only the largest of the Great Lakes, it is also the largest body of freshwater by surface area on Earth. Its surface area of 82,100 square kilometers (31,700 square miles) is greater than the combined areas of Vermont, Massachusetts, Rhode Island, Connecticut and New Hampshire. Lake Superior's water volume of 12,100 cubic kilometers (2,900 cubic miles) equals 10% of the world's surface freshwater. Although only one third of Lake Superior's surface area, Lake Baikal in Siberia holds twice as much water thanks to its extraordinary depth of 1,642 meters (5,387 feet).

13.65 billion liters (3.6 billion gallons)

The amount of water that Lake Mead has lost since 1998

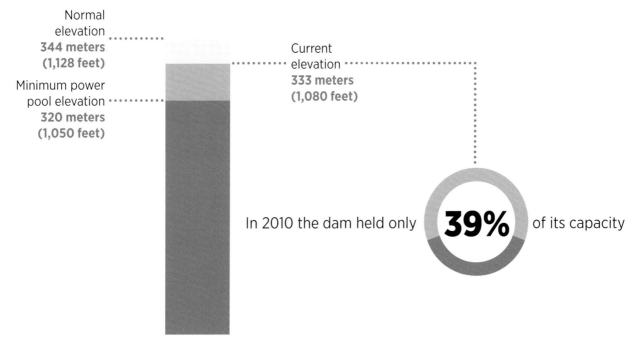

Normal elevation
344 meters (1,128 feet)

Minimum power pool elevation
320 meters (1,050 feet)

Current elevation
333 meters (1,080 feet)

In 2010 the dam held only **39%** of its capacity

Hydrologists have projected that Lake Mead has a **50% chance of running dry** by the year 2025.

Lake Mead, the largest reservoir in the United States, is 180 kilometers (112 miles) long. Formed by the construction of the Hoover Dam in the 1930s, it supplies about 85% of Las Vegas's water and 15% of its electricity. A 14-year drought and increasing consumption have led to a dangerous drop in the lake's levels. Since 1998 it is estimated to have lost 13.65 billion liters (3.6 billion gallons), its receding marked by a white "bathtub ring" of mineral deposits. If Lake Mead fell below the minimum power pool elevation of 320 meters (1,050 feet) above sea level, the Hoover Dam's turbines would shut down and the lights would start going out in Las Vegas. Hydrologists have projected that Lake Mead has a 50% chance of running dry by the year 2025.

90%
The amount the Aral Sea shrank after having its waters diverted

In the 1960s, to irrigate crops, the former Soviet Union diverted two rivers that fed the Aral Sea

In 20 years	In 30 years	In 40 years
The fishing industry was destroyed	Surface area shrank by **60%** Volume shrank by **80%**	Surface area shrank to **10%**

The Aral Sea was **halved into two small lakes** | the **equivalent of draining** Lake Erie and Lake Ontario | **Salinity increased fivefold**, killing most flora and fauna

Once the fourth largest lake in the world, the Aral Sea is viewed as one of the world's worst environmental disasters. In the 1960s, the waters of the two rivers that fed the Aral Sea were diverted to irrigate rice, melons, cereals, cotton and other crops grown in the desert. By 1998 the sea's volume had shrunk by 80%; the amount of water lost was the equivalent of completely draining Lakes Erie and Ontario. By 2007 the Aral Sea's surface area had shrunk to just 10% of its original size.

4,000,000,000,000

4 trillion Olympic-sized swimming pools would be needed to **hold all the Earth's available groundwater reserves**.

The 3 largest aquifers hold

Great Artesian Basin	**Guarani Aquifer**	**Ogallala Aquifer**

64,900 cubic kilometers (15,600 cubic miles)	40,000 cubic kilometers (9,600 cubic miles)	3,608 cubic kilometers (900 cubic miles)

The Earth's largest reserves of non-frozen freshwater lie in underground layers of water-bearing permeable rock called aquifers. The groundwater in these aquifers could fill 4 trillion Olympic-sized swimming pools. More than a million years old, the Great Artesian Basin in central and eastern Australia is one of the largest confined aquifers in the world, underlying an area of almost 2 million square kilometers (772,000 square miles). Other great aquifers include the 1.2 million square kilometer (463,000 square mile) Guarani Aquifer in South America and the 450,600 square kilometer (174,000 square mile) Ogallala Aquifer in the central United States.

The Ogallala Aquifer

The Ogallala Aquifer
3,608 cubic kilometers
(866 cubic miles)

provides **81%** of water used in the Great Plains

and is falling at a rate of **2.7 meters (9 feet) per year**

Which is **14X** faster than can be recharged naturally

At 3,608 cubic kilometers (866 cubic miles), the Ogallala Aquifer is one of the largest underground water reserves in the world. Its total water storage is close to that of Lake Huron, and it provides 81% of water used in the Great Plains. Farming accounts for 94% of the groundwater that is used to irrigate nearly 5.5 million hectares (13.6 million acres) of farmland in the Great Plains. But since large-scale irrigation began in the 1940s, water levels have declined dramatically. The Ogallala took 6,000 years to fill, but it is now being used up 14 times faster than it can be recharged naturally, sinking at a rate of 2.7 meters (9 feet) every year.

50 meters (164 feet)

The depth that aquifers beneath some major world cities have dropped

which is about **half the height** of the Statue of Liberty (from base of pedestal to top of torch)

Using too much water from an aquifer can lower the water table. In 60% of European cities with more than 100,000 people, groundwater is being extracted faster than it can be replenished. The aquifers underlying Mexico City fell 5–10 meters (16–33 feet) between 1986 and 1992. The water levels for Beijing, Madras and Shanghai have fallen 10–20 meters (33–66 feet), and those of Bangkok, Manila and Tianjin have all dropped 20–50 meters (66–164 feet)—about half the height of the Statue of Liberty.

300 meters (1,000 feet)

The depth to which some wheat farmers in northern China need to drill wells to reach sinking groundwater.

which is **nine-tenths the height** of the Eiffel Tower

Expanding cities and intensive farming have meant that water tables have dropped precipitously in the arid North China Plain, with more water being pumped out annually than is being replenished. Two-thirds of the groundwater below cities such as Shijiazhuang, which boasts more than 800 illegal wells, has been drained. Groundwater levels in the North China Plain have dropped as much as 1 meter (3 feet) per year, forcing some farmers to drill wells as deep as 300 meters (1,000 feet)—about the height of the Eiffel Tower—to find water.

1.6 million hectares

Area of global farmland abandoned **annually** because of **salt buildup** from irrigation

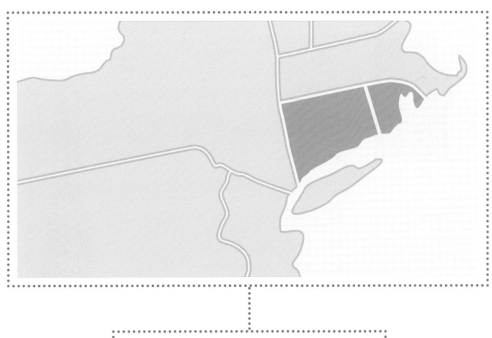

Which is roughly **an area the size of Connecticut and Rhode Island.**

All water contains dissolved mineral salts. If farmland is over-irrigated without proper drainage, salts will build up in the soil, reducing crop yields and eventually rendering the land infertile. It is estimated that salt buildup affects 20% of the world's farmland. Salt has effectively damaged 36% of all irrigated land in India. Australia is projected to lose more than 17 million hectares (42 million acres) of farmland to salt by 2050.

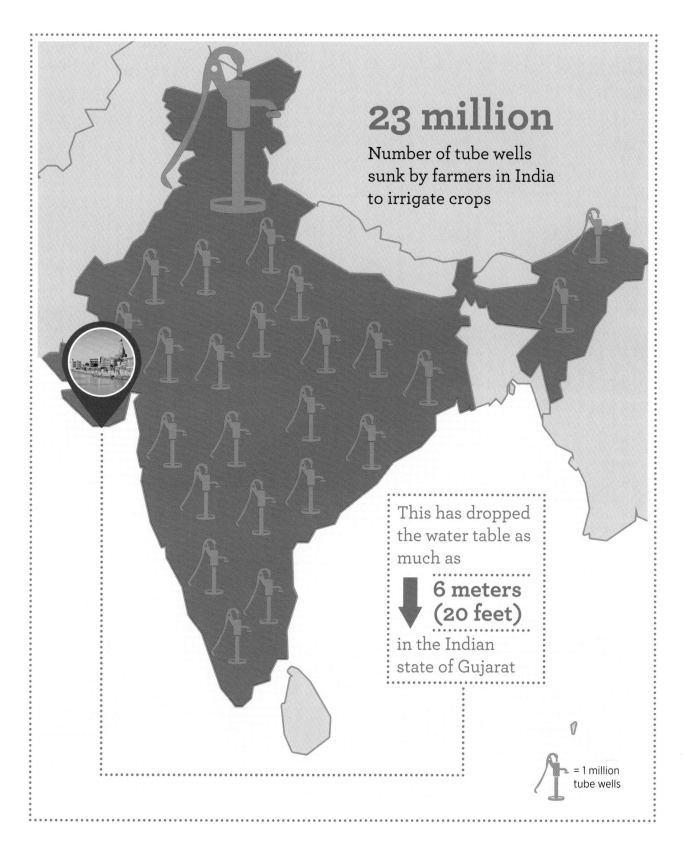

23 million

Number of tube wells sunk by farmers in India to irrigate crops

This has dropped the water table as much as

6 meters (20 feet)

in the Indian state of Gujarat

= 1 million tube wells

It is estimated that 1.5 billion people worldwide rely on groundwater for survival. But groundwater is being pumped out faster than it can be replenished, and this means that wells need to be drilled deeper and deeper to reach water. In countries such as India, where farmers have relied on subsidized energy and lax regulations to sink more than 23 million tube wells, the water table has dropped as much as 6 meters (20 feet), and areas such as Gujarat face a looming serious shortage of groundwater.

2.1 billion

Number of people worldwide living in areas facing water shortages

or **30%** of the world's population

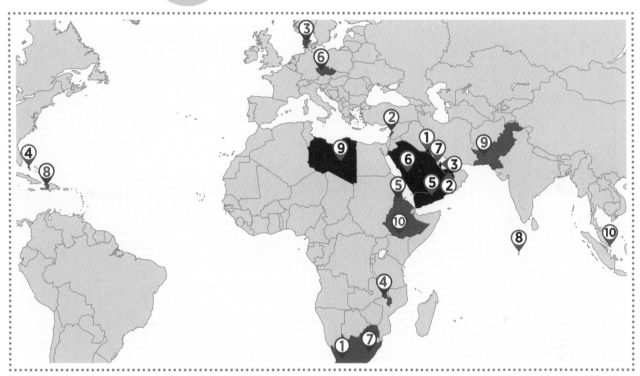

Top 10 Water-**Scarce/Stressed** Nations
Annual Freshwater Availability per Capita (cubic meters/cubic feet)

WATER-SCARCE NATIONS

Ranking	Country	Cubic Meters	Cubic Feet
1	Kuwait	7.1	250.7
2	United Arab Emirates	19.0	670.1
3	Qatar	19.0	670.1
4	Bahamas	57.6	2,034.1
5	Yemen	84.7	2,991.1
6	Saudi Arabia	85.5	3,019.4
7	Bahrain	87.6	3,093.5
8	Maldives	93.8	3,312.5
9	Libya	109.0	3,849.2
10	Singapore	115.7	4,061.1

WATER-STRESSED NATIONS

Ranking	Country	Cubic Meters	Cubic Feet
1	South Africa	1,019	35,985.6
2	Lebanon	1,057	37,327
3	Denmark	1,077	38,033
4	Malawi	1,123	39,658.3
5	Eritrea	1,163	41,070.9
6	Czech Republic	1,248	44,072.7
7	Lesotho	1,377	48,628.2
8	Haiti	1,386	48,946.1
9	Pakistan	1,396	49,299.2
10	Ethiopia	1,440	50,853.1

According to the United Nations, a country is said to experience "water stress" when annual available freshwater supplies drop below 1,700 cubic meters (60,000 cubic feet) per person. When supplies drop below 1,000 cubic meters (35,300 cubic feet) per person, the population faces "water scarcity," and below 500 cubic meters (17,660 cubic feet),

"absolute scarcity." Water-stressed regions must make painful decisions about water use for personal consumption, agriculture and industry, while water scarcity severely hinders economic development, strains the environment and limits food availability. Some 47 countries face water shortages, 18 are water-stressed and 29 are classified as water-scarce.

Projections for the year 2050

The world population is predicted to grow from **6.9 billion** in 2010 to **9.1 billion**

The world's urban population is forecast to grow from **3.4 billion** in 2009 to **6.3 billion**

Current food demand

2050 food demand

4.8 billion people are projected to lack adequate freshwater

Energy demand from hydropower and other renewable energy resources **will rise by 60%**

Energy consumption is expected to **surge 58%**

Water consumption for energy is expected to **increase 11% per year**

The intersecting pressures of food production, industrialization, population growth, climate change, global conflicts, sociopolitical shifts and an expanding middle class means that water scarcity levels will increase greatly within the next half-century.

Africa's population will **grow from about 1.1 billion in 2013 to 2.4 billion**

7.5 billion people will be living in low and middle income countries, with 2 billion in sub-Saharan Africa and 2.2 billion in South Asia

These growth projections, combined with changing diets, result in a predicted increase in food demand of **70%**

The number of countries facing water stress or scarcity could rise to 54, **with a combined population of four billion people**

Global agricultural water consumption (including both rainfed and irrigated agriculture) is estimated to **increase by about 19% to 8,515 cubic kilometers per year**

Global temperatures are expected to rise by between **2° and 3°C (35.6°F and 37.4°F).** The cost of adapting to this rise could range from **$70 billion to $100 billion annually**

As a result of climate change, deforestation, loss of wetlands, rising sea levels and population growth in floodprone lands, **the number of people vulnerable to flood disasters worldwide is expected to mushroom to two billion**

Freshwater consumption to meet social and economic demands is **expected to grow 24%**

The number of people living in water-stressed basins is **expected to exceed 5 billion**

The 10 biggest American cities running out of water

RANKED BY POPULATION

5. SF Bay Area*

7. Las Vegas

3. Phoenix

9. Atlanta

8. Tucson

6. Fort Worth

1. Los Angeles

4. San Antonio

2. Houston

10. Orlando

*Note: SF Bay Area includes San Francisco, Oakland and San Jose

Some of America's largest cities are in danger of running out of water because of a chronic combination of drought, population growth and industrial thirst for water. In the short term, San Antonio and Orlando face immediate shortages.

Los Angeles, Las Vegas and Atlanta have all faced severe water shortages in the past and are projected to do so again. Houston and Tucson have been identified as the cities with the highest risk of water shortage.

Top 10 Freshwater Extracting Countries

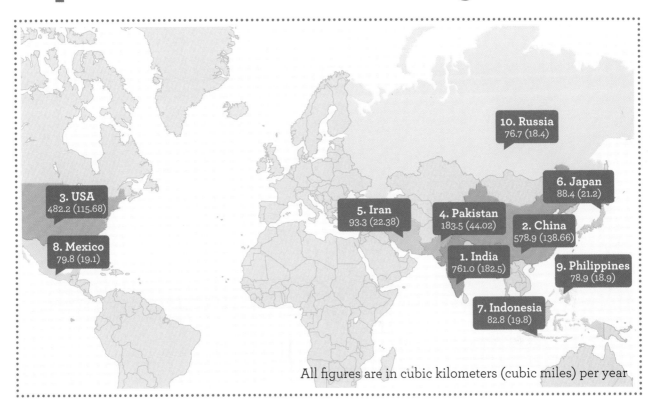

10. Russia
76.7 (18.4)

6. Japan
88.4 (21.2)

3. USA
482.2 (115.68)

5. Iran
93.3 (22.38)

4. Pakistan
183.5 (44.02)

2. China
578.9 (138.66)

8. Mexico
79.8 (19.1)

1. India
761.0 (182.5)

9. Philippines
78.9 (18.9)

7. Indonesia
82.8 (19.8)

All figures are in cubic kilometers (cubic miles) per year

India uses **761 cubic kilometers (182.5 cubic miles)** of freshwater annually

Lake Ontario

the equivalent of **draining half of Lake Ontario**

India uses more freshwater than any other country in the world, exceeding even China. In a single year India uses the equivalent of half of Lake Ontario.

90% of freshwater used in India is destined for agriculture. Freshwater withdrawals have tripled over the past 50 years, with demand increasing each year.

113,000,000,000,000 liters (30 trillion gallons)

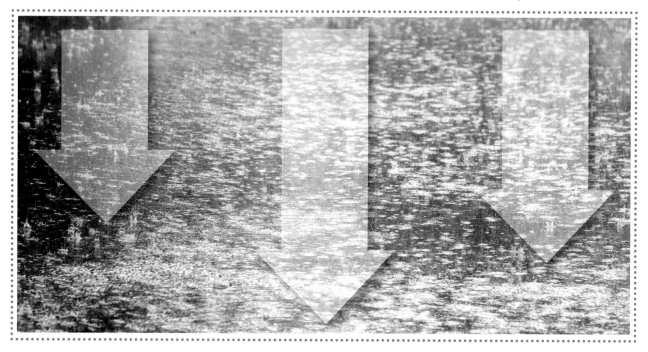

Amount of water that falls globally every day as precipitation

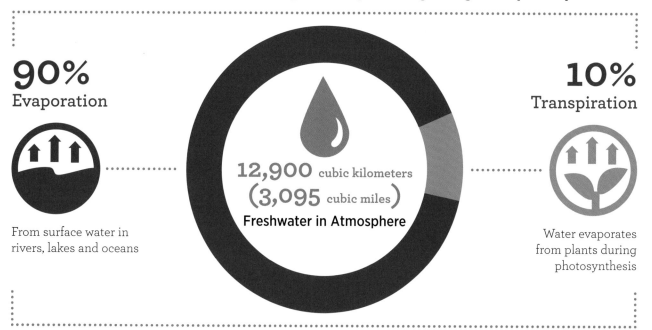

90%
Evaporation

From surface water in rivers, lakes and oceans

12,900 cubic kilometers
(**3,095** cubic miles)
Freshwater in Atmosphere

10%
Transpiration

Water evaporates from plants during photosynthesis

90% of the freshwater in the atmosphere is due to evaporation from water bodies and from photosynthesis. Condensation in the atmosphere produces clouds of water vapor that eventually release precipitation. An estimated 113 trillion liters (30 trillion gallons) of precipitation falls to Earth daily. In 2014 a new satellite was launched to provide the next-generation global observations of rain and snow as part of the Global Precipitation Measurement (GPM) mission.

Less than 100 millimeters (4 inches)

Annual precipitation in Antarctica's McMurdo Dry Valleys, one of the driest places on earth

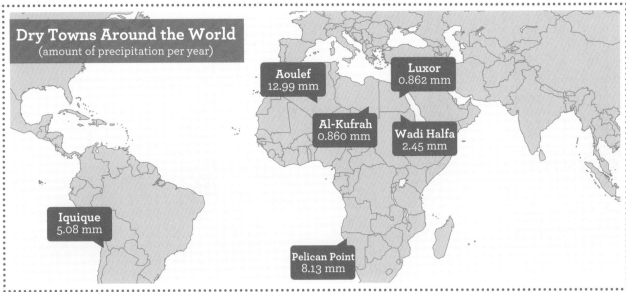

Dry Towns Around the World
(amount of precipitation per year)

Aoulef
12.99 mm

Luxor
0.862 mm

Al-Kufrah
0.860 mm

Wadi Halfa
2.45 mm

Iquique
5.08 mm

Pelican Point
8.13 mm

You would think one of the driest places on Earth would be a scorching hot desert, but it is actually at the South Pole, one of the coldest places on the planet. Blocked by high mountains, Antarctica's McMurdo Dry Valleys have extremely low humidity and almost no ice or snow cover. Another extremely arid region is the Atacama Desert in Chile, where some areas have not recorded rainfall in 400 years. Because of their dryness, both areas are considered the closest of any of Earth's environments to Mars.

26.5 meters (87 feet)

Rainfall recorded in Cherrapunji, India, "the wettest place on Earth," between August 1, 1860, and July 31, 1861

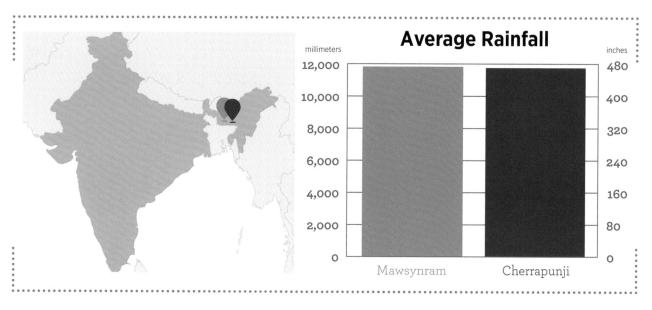

Average Rainfall

millimeters | inches

Mawsynram | Cherrapunji

Cherrapunji, India, holds the record for the most rainfall in one year—26.5 meters (87 feet) of rain between August 1860 and July 1861. Its average annual rainfall, 11.78 meters (38.65 feet), places it a close second to Mawsynram, a mere 16 kilometers (10 miles) away, which receives an average of 11.87 meters (38.94 feet) annually. The two Indian towns regularly trade the title of "wettest place on Earth."

12 meters (40 feet)

The amount Chile's Echaurren Glacier has been shrinking per year since 1970

It is forecasted to **melt entirely by 2058**

In 2009 a Chilean government study found that

92% of the country's glaciers were receding because of glacial melt caused by global warming

Of the **100** glaciers that were examined, only **7** were in stable condition.

In 2009, a study determined that 92% of Chile's glaciers were receding because of glacial melt caused by global warming. The study found that only seven of the country's glaciers were in stable condition. On the western slopes of the Andes, the Echaurren Glacier is located just 50 kilometers (31 miles) east of Santiago, Chile's capital. It supplies 70% of the city's drinking water and is shrinking at a rate of about 12 meters (40 feet) a year. It is forecast to melt entirely by 2058.

"Water, like religion and ideology, has the power to move millions of people."

—Mikhail Gorbachev

At Home

Based on the use of water in our homes we must believe that a limitless resource flows through our pipes. The average North American uses 378 liters (100 gallons) a day for bathing, washing, cooking and cleaning. When you add in water used for lawn watering, car washing and to fill swimming pools, North Americans live as though water shortages didn't exist. By contrast, in developing countries in Africa and Asia the nearest source of freshwater can be 6 kilometers (3.6 miles) away, requiring journeys of up to 6 hours a day by foot to retrieve subsistence supplies of water from a source that may be contaminated or polluted. Water scarcity—or water wealth, if you will—is determined by geography, some countries simply have much more water than others. But that doesn't mean the water-rich are entitled to squander their water. Being water conscious at home and through the purchases we make can dramatically reduce the actual water we use every day and the virtual water in our products. For example, we flush 30 percent of the water we use down the toilet each day: installing low-flow toilets, taking shorter showers and turning the tap off while brushing our teeth would go a long way to being water conscious.

The average distance women in rural Africa and Asia walk to collect water

10,000 steps,
or 6 kilometers (3.6 miles)

In developing countries in Africa and Asia, the distance to the nearest source of freshwater is typically about 10,000 steps—or 6 kilometers (3.6 miles). In Africa 90% of collecting freshwater for families is done by women, who on average spend about six hours a day making several trips while carrying heavy containers on their heads.

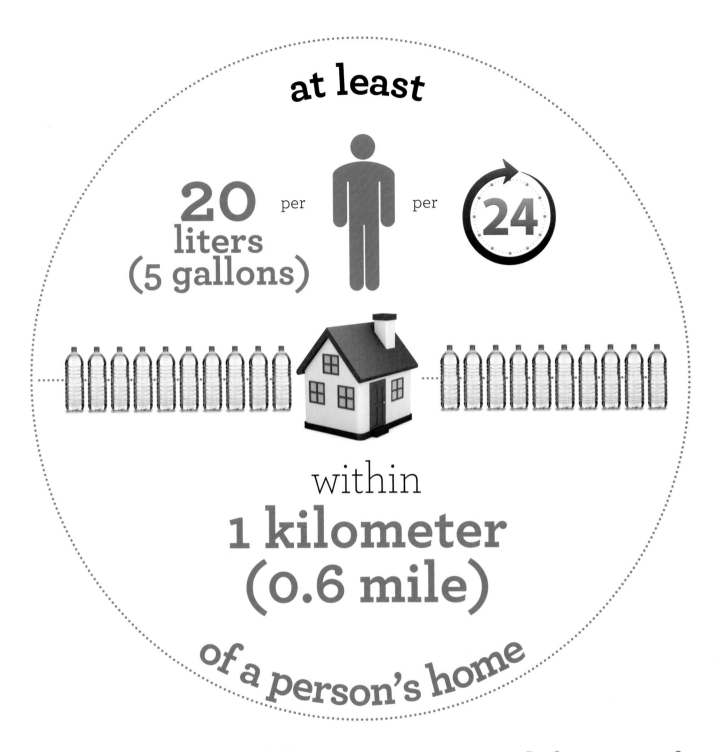

at least

20 liters (5 gallons) per per 24

within 1 kilometer (0.6 mile)

of a person's home

The **World Health Organization's** definition of what constitutes **"reasonable access to water"**

"Reasonable access" to water means having a source of at least 20 liters (5 gallons) per person per day within 1 kilometer (0.6 mile) of the water user's home. In the United States and Canada, the average person uses 375 liters (100 gallons) each day for household activities and yardwork, such as drinking, cooking, washing and gardening.

Real vs. Synthetic leather

Leather Sofa

A 3-seater leather sofa requires

8 kilograms (18 pounds) of leather

Feeding cattle and tanning their hides requires

X 944

17,000 liters per kilogram
(4,500 gallons for 2 pounds) of leather

Total Water Usage

**136,000 liters
(36,000 gallons)**

Feeding cattle and tanning their hides drives the virtual water footprint of leather to 17,000 liters per kilogram (4,500 gallons for 2 pounds). Covering a three-seater couch requires 8 kilograms (18 pounds) of leather, which works out to a water footprint of 136,300 liters (36,000 gallons) of water.

Synthetic Sofa

A 3-seater synthetic sofa requires

23 meters (25 yards) of fabric

Synthethic sofa fabric requires

X 4

70 liters (18.5 gallons) for
1 kilogram (2 pounds) of fabric

Total Water Usage

**1,610 liters
(425 gallons)**

By comparison, it takes only 70 liters (18.5 gallons) of water to produce 1 meter (3 feet) of a synthetic sofa fabric such as rayon or vinyl. The same sofa could be covered for a fraction of the water used—1,610 liters (425 gallons).

The water footprint of **1 cloth diaper** is

15 liters (4 gallons)

The water footprint of **1 disposable diaper** is

545 liters (144 gallons)

To grow the cotton needed to produce one cloth diaper requires 750 liters (198 gallons) of water; yet because the diaper will be reused an average of 50 times its water footprint is just 15 liters (24 gallons). Disposable diapers, because they are made of fluffed wood pulp, superabsorbent polymer and non-woven polypropylene, require a whopping 545 liters (144 gallons) each. Parents who opt for disposable diapers will go through about 8,000 diapers, with a water footprint of close to 4.4 million liters (1.2 million gallons) of water.

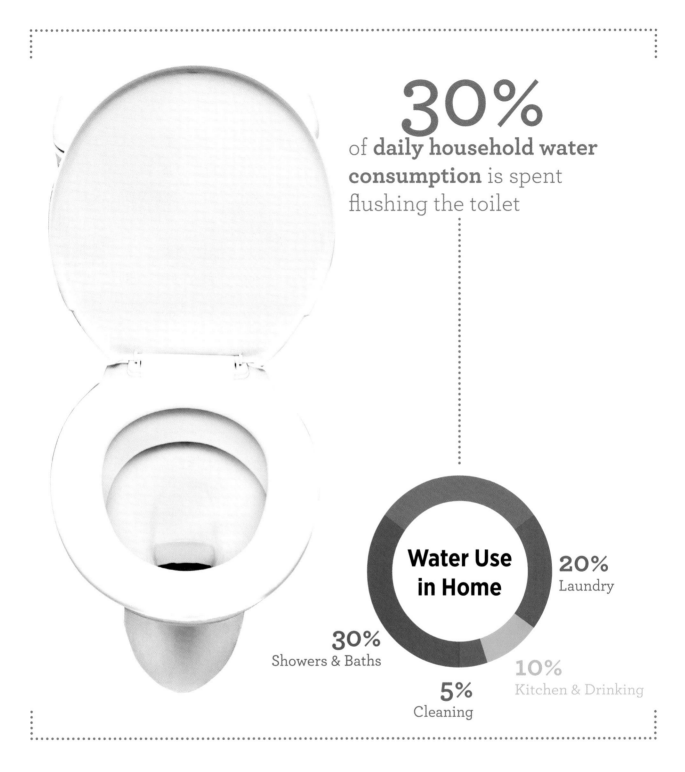

30%

of **daily household water consumption** is spent flushing the toilet

Water Use in Home

20%
Laundry

30%
Showers & Baths

5%
Cleaning

10%
Kitchen & Drinking

The average domestic toilet is flushed 5 times daily per person. High efficiency toilets use as little as 3.75 liters (1 gallon) per flush, whereas the standard for older toilets is 16 to 23 liters (3.5 to 5 gallons) per flush. Using older toilets, the average person flushes 80 to 100 liters (21 to 26 gallons) of high-quality drinking water down the pipes every day. That amounts to 30% of an individual's daily water consumption.

To produce **1 kilowatt hour**
of electricity requires

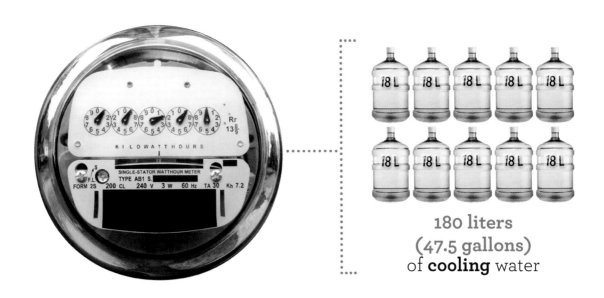

180 liters
(47.5 gallons)
of **cooling** water

One kilowatt hour is enough power to

Light a 100 w incandescent bulb for 10 hours

Surf the web for 5 hours

Dry your hair 3 times

Bake 1 birthday cake

Thermoelectric power plants use steam-driven turbines to generate electricity. Whether fossil fuel or a nuclear reaction is used as the heat source, water is needed both to make the steam and to cool it back into water once it has passed through the turbines. While the water is typically returned to its natural source once cooled, excessive withdrawals can result in thermal pollution and damage to local ecosystems.

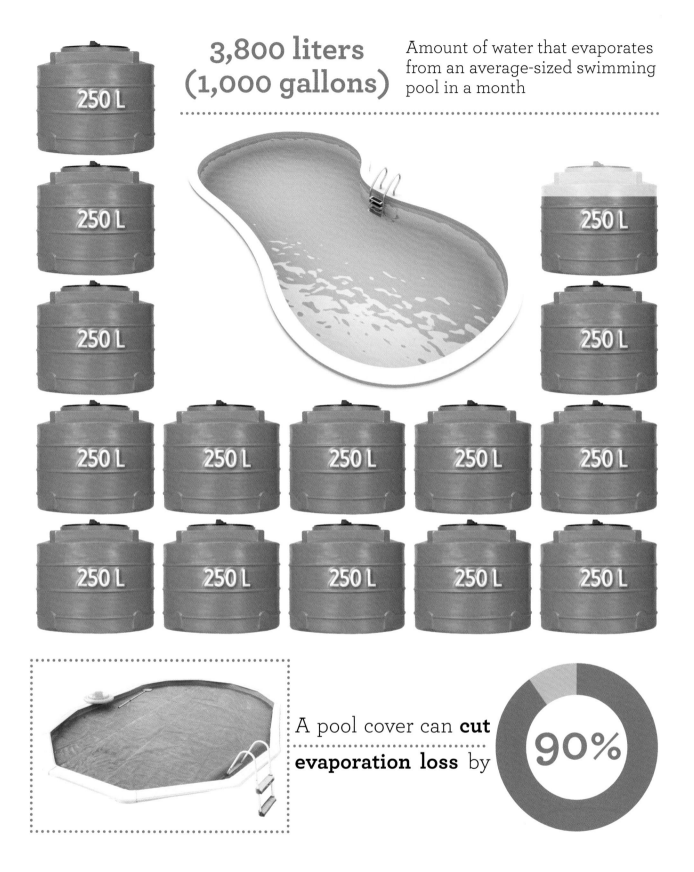

3,800 liters (1,000 gallons)

Amount of water that evaporates from an average-sized swimming pool in a month

250 L

250 L

250 L

250 L

250 L

250 L

250 L

250 L

250 L

250 L

250 L

250 L

250 L

250 L

250 L

250 L

A pool cover can **cut evaporation loss** by **90%**

There are more than 5 million homes in the United States (and more than 1 million in Canada) with swimming pools. The average swimming pool holds 50,000 liters (13,000 gallons) of water and loses approximately 3,800 liters (1,000 gallons) per month to evaporation. Heating the pool can make the water evaporate even faster; a pool cover can cut this loss by up to 90%.

8,000 liters (2,100 gallons)

Amount of water used **per round of golf** to maintain courses in dry climates

Keeping the world's golf courses lush and green takes an estimated **9.4 billion liters (2.5 billion gallons) of water** every single day

which is about **four times** the volume of the Great Pyramid of Giza

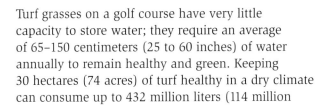

Turf grasses on a golf course have very little capacity to store water; they require an average of 65–150 centimeters (25 to 60 inches) of water annually to remain healthy and green. Keeping 30 hectares (74 acres) of turf healthy in a dry climate can consume up to 432 million liters (114 million gallons) of water annually, which works out to around 8,000 liters (2,100 gallons) per round of golf. It is estimated that it takes 9.4 billion liters (2.5 billion gallons) of water to irrigate the world's golf courses every day.

Average **residential water use** by **Canadians**

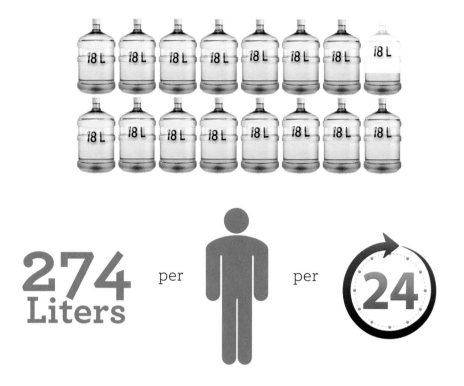

274 Liters per 👤 per 🔄 **24**

Residential **Water Use** by Province

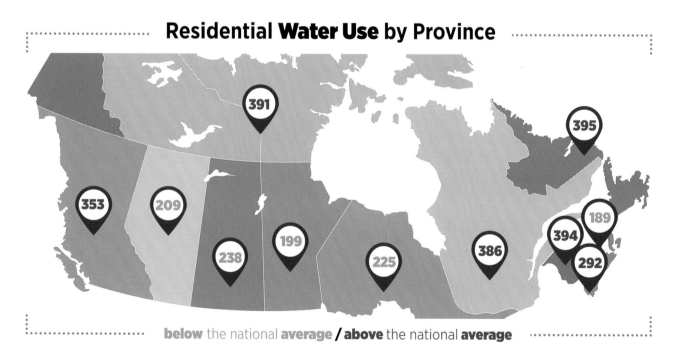

below the national **average** / **above** the national **average**

In 2009, Canadians consumed an average of 275 liters (72 gallons) of water per capita per day for residential use. Water usage varied by province, with the largest residential use being found in Quebec and the Maritimes and the smallest in the Prairie Provinces.

Average **Residential Water Use by Americans**

Top 5 Highest Liters (Gallons) per Day

719 (190)	708 (187)	704 (186)	625 (165)	575 (152)
Nevada	Idaho	Utah	Hawaii	Wyoming

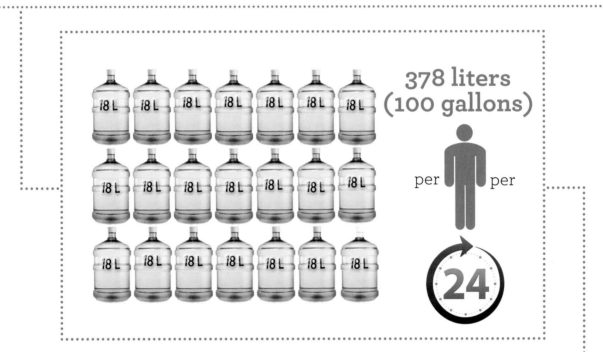

18 L 18 L 18 L 18 L 18 L 18 L 18 L
18 L 18 L 18 L 18 L 18 L 18 L 18 L
18 L 18 L 18 L 18 L 18 L 18 L 18 L

378 liters (100 gallons)

per / per

24

Top 5 Lowest Liters (Gallons) per Day

204 (54)	216 (57)	216 (57)	231 (61)	241 (64)
Maine	Wisconsin	Pennsylvania	Delaware	Vermont

Americans consume an average of 378 liters (100 gallons) of water per day per person. In general, domestic water use rates are lowest in the northern and eastern states, whereas the highest levels of consumption are found in the drier mountain and western states.

16.9 %

Growth rate of bottled water consumption in Thailand between 2006 and 2011, outpacing China and **more than three times the world average of 5.5%**

Top 10 **Countries Consuming Bottled Water** & Their Growth Rate (2006–2011)

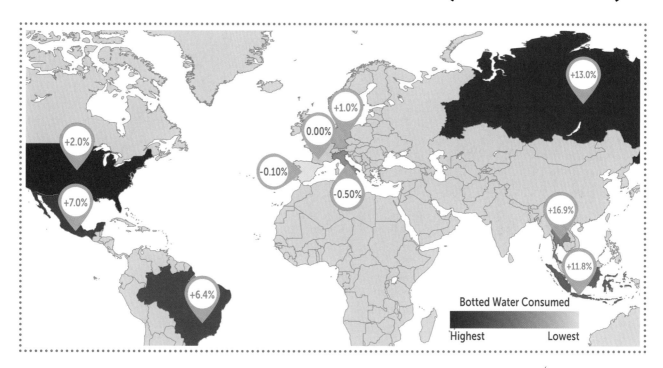

+13.0%

+1.0%

0.00%

+2.0%

-0.10%

-0.50%

+7.0%

+16.9%

+11.8%

+6.4%

Botted Water Consumed

Highest   Lowest

The mania for bottled water continues to grow around the world. The United States leads the way in annual purchases, consuming 38 billion liters (10 billion gallons) in 2013. But China and Southeast Asia are growing the fastest. In 2009, Americans spent $21 billion on bottled water, almost as much as the $29 billion spent to maintain America's water systems.

A **10-minute shower** uses

160–190 liters (40–50 gallons) of water

Getting clean in less than four minutes, using a water-efficient showerhead, **reduces use by 100+ liters.**

Most people think that showers use less water than baths, but research done in the United Kingdom proves just the opposite. Leisurely showers can easily drain more than 160 liters (42 gallons), while baths average around 75–80 liters (20–21 gallons).

A 10-minute shower uses approximately 160–190 liters (42–50 gallons) of water. Being expeditious and getting clean in less than 4 minutes, using a water-efficient showerhead, cuts use levels by more than 100 liters (26 gallons).

A five-minute shave with running water uses

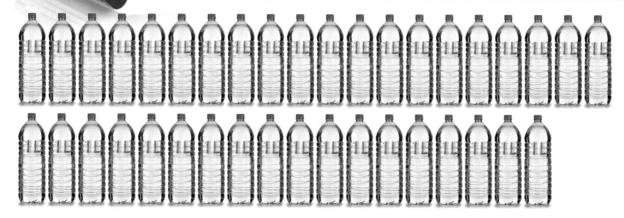

—— 38 liters (10 gallons) of water ——

For a man who shaves daily, that's a whopping **13,870 liters (3,650 gallons) a year,** or the equivalent of a mid-sized water tanker.

The average bathroom faucet runs at 7.6 liters (2 gallons) per minute. A typical five-minute shave with the water running entails about 38 liters (10 gallons) of water going down the drain.

For men who shave daily, that's 13,870 liters (3,650 gallons) a year, or the equivalent of a mid-sized water tanker. Simply turning off the tap and filling the sink can cut usage by two-thirds.

To produce
1 cotton shirt
requires

To produce
1 polyester shirt
requires

**350 liters
(92.5 gallons)
of water**

**2,500 liters
(660 gallons)
of water**

Nearly 11,000 liters (2,900 gallons) of water is needed to produce 1 kilogram (2 pounds) of printed cotton textile. Thus a 250-gram (9-ounce) cotton shirt has a water footprint of 2,750 liters (725 gallons). Of this total water volume, 33% is blue irrigation water used by the cotton plant, 54% is green rainwater and 14% is gray water used in the processing operations and washings (see p. 72 for definitions). Polyester has a nearly zero water footprint, because manufacturing the polymer uses very little water. However, the creation of polyester cloth uses the same dyestuffs, finishing chemicals and washes as cotton, giving a polyester shirt a water footprint of around 350 liters (92 gallons).

To produce **1 pair** of jeans requires

**7,600 liters
(2,000 gallons)
of water**

················· *This includes* ·················

Growing Cotton   +   Manufacturing

It takes more than 7,600 liters (2,000 gallons) to make a typical pair of jeans. This does not include water used in laundering them over their lifetime, which would produce a more staggering result.

To produce **1** integrated circuit board measuring 30 centimeters (12 inches) requires ······················

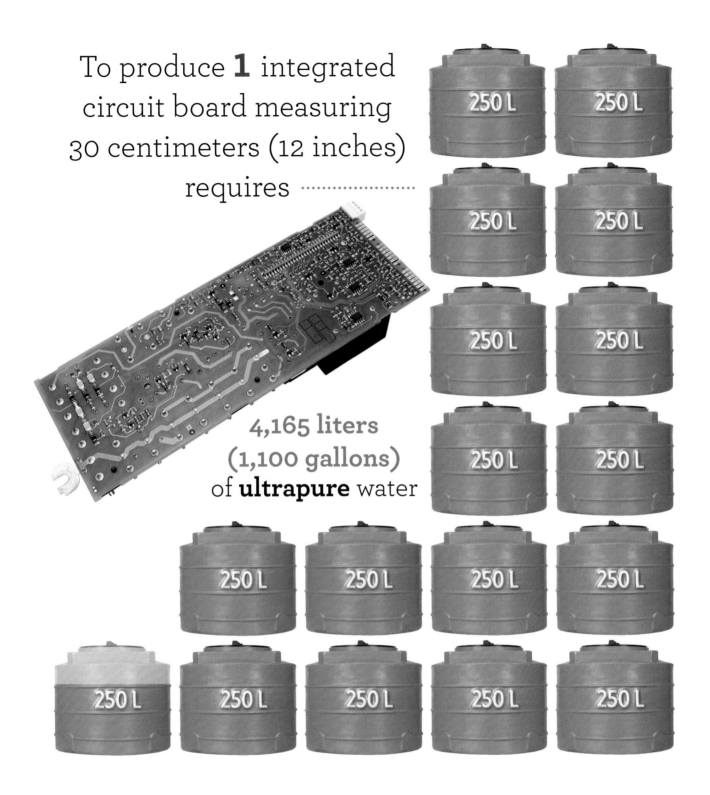

4,165 liters (1,100 gallons) of **ultrapure** water

To make microchips requires lots of water. Silicon semiconductors with integrated circuits must be scrubbed free of debris with the cleanest water possible. Ultrapure water (UPW), which is 10 million times cleaner than regular water, requires 12 filtration steps. One IBM microchip plant in Burlington, Vermont, produces 7.5 million liters (2 million gallons) of UPW a day.

To produce **1** smartphone requires

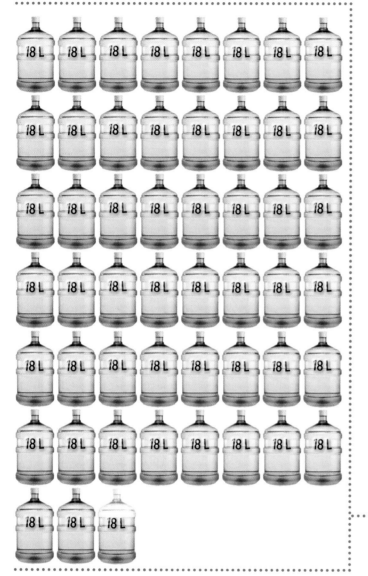

910 liters (240 gallons)
of water

Cellphones and smartphones use water throughout their production process, from creating the microchips to mining the metals used in the batteries to polishing the silica glass used in their touch screens. In total, each phone requires about 910 liters (240 gallons) of water to manufacture. The number of activated cellphones is soon expected to exceed the world's population. To manufacture these phones will require 6.7 trillion liters (1.8 trillion gallons) of water, much of it blue and gray (see p. 72).

A **single pair** of
leather shoes uses

.... **8,000 liters**
(2,113 gallons)
of water ....

6 kilograms
(13 pounds)
of leather

17,000 liters
of water per kilogram
(2,100 gallons per pound)

550 kilograms
(1,200 pounds)

Most leather shoes are made from cowhide. A beef cow at the end of its life will have a water footprint of around 1.9 million liters (500,000 gallons). On average, 5.5% of this water is attributed to the cow's hide. A beef cow will produce about 6 kilograms (13 pounds) of leather, so the water footprint of bovine leather is 17,000 liters per kilogram (2,100 gallons per pound). The tanning process also involves many water-intensive processes. At 500 grams (1 pound) of leather for a pair of shoes, this works out to about 8,000 liters (2,113 gallons) of water for those new wingtips. Alternatively, suede made from lambskin can produce a pair of shoes that uses one-third of the water of cow leather—2,880 liters (760 gallons).

A single rose stem requires

9.2 liters (2.4 gallons) of water

215 million roses are sold in the U.S. every Valentine's Day

That requires almost **2 billion liters (528 million gallons) of water**

Many of the cut flowers we buy originate in Kenya, particularly the area around Lake Naivasha, in the Rift Valley northwest of Nairobi. Cut flowers are Kenya's third largest national industry after tea and tourism. The water footprint of cut-flower production around Lake Naivasha has more than doubled over the past 15 years. A 25-gram (1-ounce) rose stem takes 9.2 liters (2.4 gallons) of water to grow and harvest. In the United States, 215 million roses are sold every Valentine's Day, adding up to almost 2 billion liters (528.3 million gallons) of water.

To produce a **1 carat** diamond requires

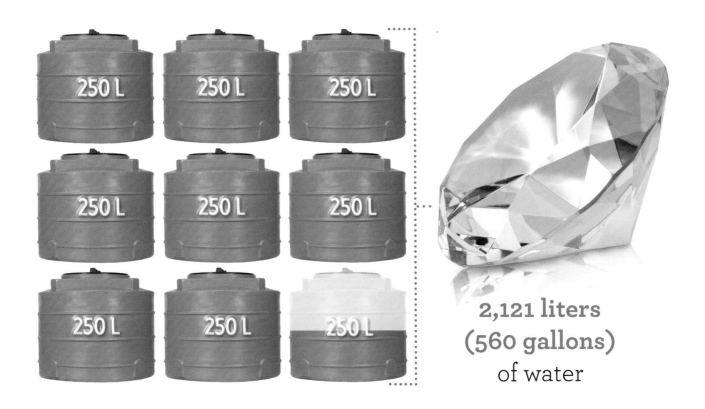

250 L · 250 L · 250 L
250 L · 250 L · 250 L
250 L · 250 L · 250 L

**2,121 liters
(560 gallons)
of water**

Essential vs.
Aesthetic Value

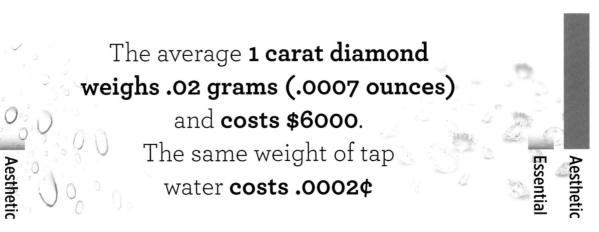

The average **1 carat diamond
weighs .02 grams (.0007 ounces)**
and **costs $6000**.
The same weight of tap
water **costs .0002¢**

Essential Aesthetic   Essential Aesthetic

Something as essential to life as water is valued less in the marketplace than diamonds, which are nonessential but have a high aesthetic and status value. Tap water sells for pennies per gallon while a 1-carat diamond fetches about $6,000. Ounce for ounce, the price of diamonds is billions of times that of water. A lot of water is used in blasting diamonds from the earth and scrubbing and washing the crushed ore: in 2010 it took 2,121 liters (560 gallons) of water to produce a 1-carat diamond.

"Saving water at home is good, but since 86% of humanity's water footprint is not within people's homes but in making food, natural fibers, oils and energy, it is crucial to consider what you buy as well."

—Professor Arjen Hoekstra

Food

All forms of food, whether plant or animal, require water to produce. Putting a bag of groceries on the table is no simple task when water is factored in to the mix. A meat-based diet requires more than twice as much water as a vegetarian-based diet, and coffee beans grown in far-off lands require more water than growing tea leaves on a per-cup basis. When vegetables grown in a water-stressed region are exported to a water-rich region, the water-stressed region is effectively exporting a scarce resource embodied in the food being exported. In a global economy, there is no clear right or wrong decision when choosing between different food products—especially when nutrition, price and seasonality are added into the equation. And depending on where food is produced, the water footprint may be big or small. The complicated decision of what foods to buy based on how much water is used can only be based on general guidelines: a vegetarian diet uses less water than a meat-based one, buy locally produced food rather than imported products, and consider water availability in the place where the food was grown or produced when selecting products from far away.

Guideline
Breakdown

chemical
contaminants

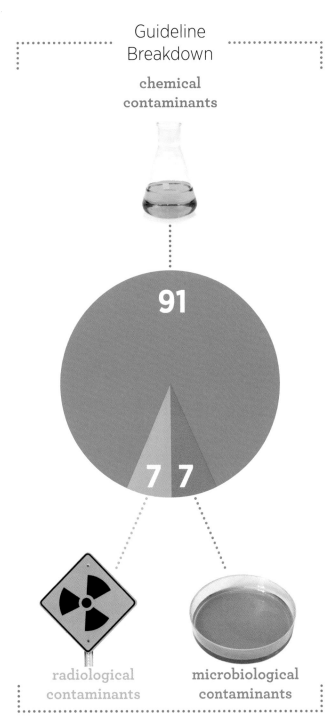

91

7 7

radiological
contaminants

microbiological
contaminants

While Canadian drinking water supplies are generally of excellent quality, as of 2013 there were 105 separate guidelines for testing drinking water that dealt with microbiological (7), chemical (91) and radiological (7) contaminants. The highest-priority guidelines are those that deal with microbiological contaminants—such as bacteria, protozoa and viruses such as *E. coli* and *Cryptosporidium*— although the water is tested for everything from mercury, gasoline, cyanide, asbestos and trichloroethylene to cesium, lead and strontium 90.

The number of different contaminants identified in American drinking water in 45 states between 2004 and 2009:

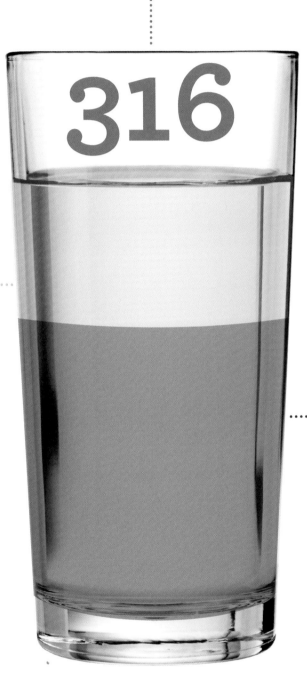

114
of these contaminants **were within Environmental Protection Agency safety standards**

316

202
chemicals were found that **are not subject to any government regulations** for drinking water

A study of water quality between 2004 and 2009 revealed agricultural pollutants, industrial chemicals and wastewater byproducts (including nitrates, arsenic and strontium 90) in the water supplied to 256 million Americans. While 114 of those contaminants were within Environmental Protection Agency safety standards, 202 chemicals were found that are not subject to any government regulations for drinking water. The cleanest tap water was found to be in Arlington, Texas, and the poorest in Pensacola, Florida, where 45 contaminants and 35 industrial pollutants were detected.

To produce **1 cup** of tea requires

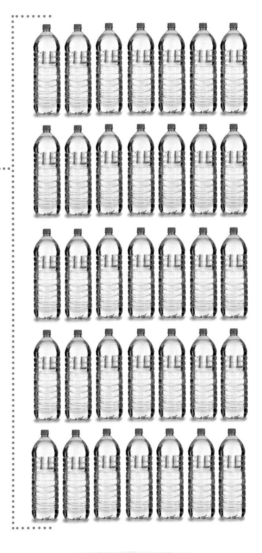

35 liters (9 gallons) of water

The virtual water content of tea consists mainly of rainwater. The dried black tea leaves needed for 1 cup (8 fluid ounces) of tea using a single teabag requires around 35 liters (9 gallons) of water to grow, process and brew. If you take lemon, milk or sugar with your tea, your cup's water footprint will be even higher—to grow the sugar cane, lemon trees and grass used to feed the dairy cow.

To produce **1 cup** of coffee requires

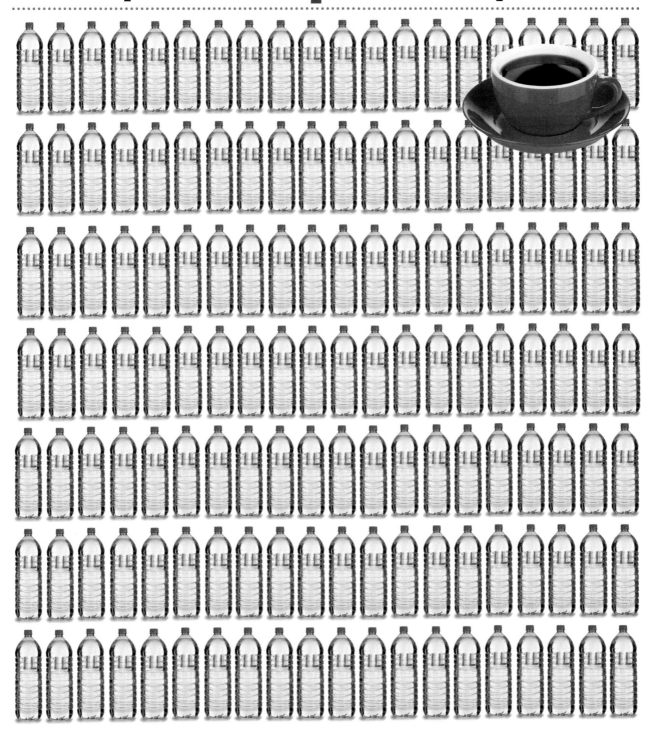

140 liters (37 gallons) of water

Making coffee is one of the largest uses of drinking water in North America. One cup (8 fluid ounces) of coffee requires 140 liters (37 gallons) of virtual water to get the beans to your grinder. By comparison, a cup of tea uses only 35 liters (9 gallons) of virtual water.

To grow **1 apple** (150 grams/5 ounces) requires

125 liters (33 gallons) of water

Water requirements for other fruits

Mango	Avocado	Pineapple
310 liters (82 gallons)	161 liters (42.5 gallons)	130 liters (34 gallons)

Water footprints can be broken down into three components: a **blue water footprint**, indicating the volume of surface and groundwater consumed during the production; a **green water footprint**, indicating the volume of rainwater consumed; and a **gray water footprint**, indicating the amount of water required to restore water purity. Fruits naturally have a high green water footprint; for example, 68% of an apple's water footprint is green.

1 watermelon (5 kilograms/11 pounds) requires

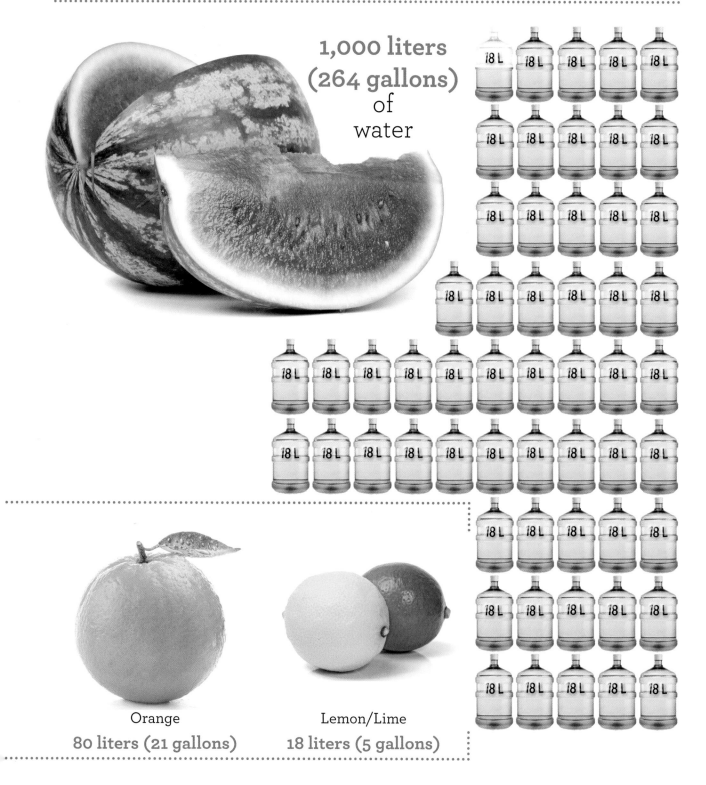

1,000 liters (264 gallons) of water

Orange
80 liters (21 gallons)

Lemon/Lime
18 liters (5 gallons)

At 92% water, a watermelon naturally requires a lot of water to grow—it needs constant watering or irrigation during the crop season. A 5-kilogram (11-pound) Green Giant or Crimson Sweet will soak up more than 1,000 liters (264 gallons) of water during its growth, harvesting and distribution.

To produce **1 kilogram (2 pounds)** of tomatoes requires

214 liters (56.5 gallons) of water on average

Tomato products require even more water

1 KILOGRAM (35 OUNCES) KETCHUP

530 liters (140 gallons) of water

The tomato is one of the world's most important foods; around 115 million tonnes (127 million tons) of the fresh fruit are produced every year. Tomatoes need a controlled supply of water throughout their 90- to 120-day growing period. On average, this puts the tomato's global water footprint at around 214 liters (56.5 gallons) for a 1-kilogram (2-pound) basket—50% of this is green water, 30% blue water and 20% gray water. Tomato products require even more water; producing ketchup, for instance, more than doubles the amount required.

To produce **1 kilogram (2 pounds)** of bananas requires

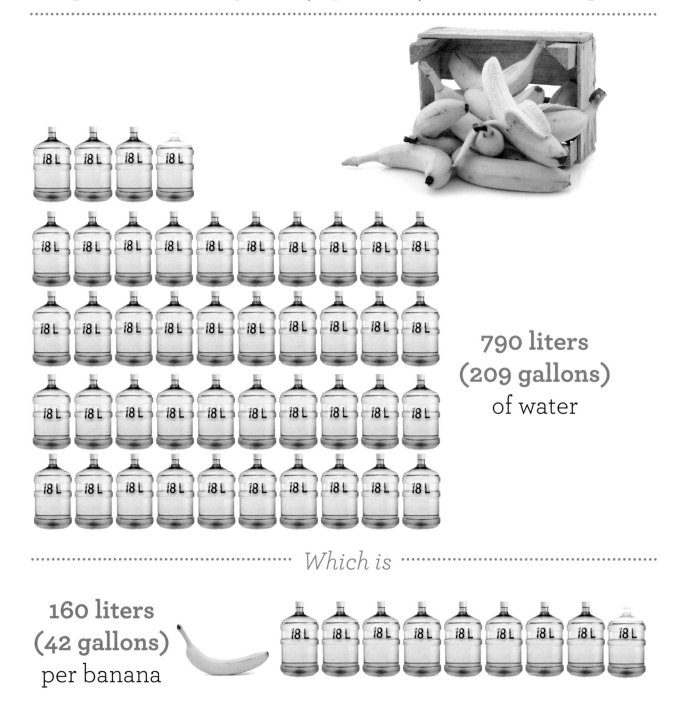

790 liters
(209 gallons)
of water

Which is

160 liters
(42 gallons)
per banana

The banana is one of the world's most popular fruits—Americans eat more bananas yearly than apples and oranges combined. To grow and process bananas takes about 790 liters (209 gallons) of water for every kilogram (2 pounds), or about 160 liters (42 gallons) per banana. Growing the fruit requires about 84% of this, but once picked, bananas go through several washings in huge tanks of freshwater.

3.1 million liters (820,000 gallons)

Amount of water required to produce 200 kilograms (441 pounds) of boneless beef

To raise a cow requires

Drinking Water	Roughage	Servicing Water	Grains
24,000 liters	**7,200 kilograms**	**7,000 liters**	**1,300 kilograms**
(6,340 gallons)	**(15,873 pounds)**	**(1,849 gallons)**	**(2,865 pounds)**
	(pasture, dry hay, silage)		(wheat, oats, barley, corn, dry peas, soybeans)

Animals have a much larger water footprint than crops. Cattle, whether raised for meat or for dairy, have an average water footprint of 15,400 liters per kilogram of beef. The lion's share (83%) of this water is attributed to the derived beef, 5.5% to the hide, and the remaining water to other beef byproducts such as the carcass, offal and semen. The global water footprint of beef production between 1996 and 2006 was about one-third of the total water footprint of all animal production.

To produce **1 kilogram (2 pounds)** of beef requires

15,400 liters (4,068 gallons) of water

That's **almost 1.5 times the volume** of a **concrete mixer truck**

Other meats require

Lamb	Pork	Goat	Chicken
10,400 liters per kilogram	**5,990 liters per kilogram**	**5,500 liters per kilogram**	**4,300 liters per kilogram**
(2,747 gallons per 2 pounds)	(1,582 gallons per 2 pounds)	(1,453 gallons per 2 pounds)	(1,136 gallons per 2 pounds)

The global water footprint of beef production is around 800 billion cubic meters (192 cubic miles) per year. Animal products almost always have a larger water footprint than crops because of the massive amounts of feed needed to nourish livestock.

Feed crops account for a whopping 99 % of beef's water footprint. The actual water footprint of beef depends strongly on how the cow is raised and the composition and origin of its feed.

A **meat-based** diet

250 liters/66 gallons

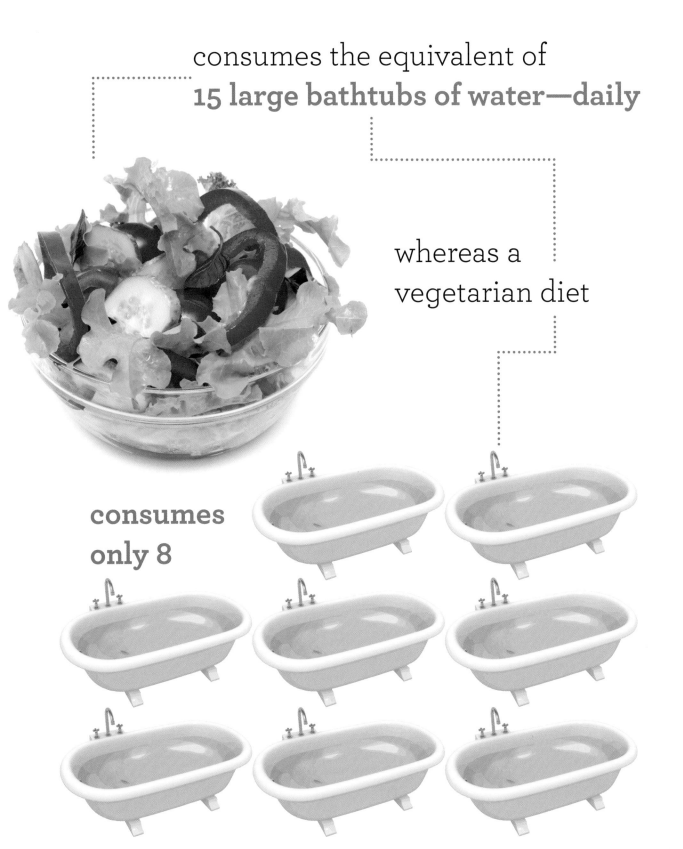

consumes the equivalent of
15 large bathtubs of water—daily

whereas a
vegetarian diet

**consumes
only 8**

In an industrialized country, a meat-based diet of 3,400 calories per day consumes between 3,600 and 5,000 liters (950–1,320 gallons) of virtual water daily. This is enough water to fill a large bathtub 15 times. Though it provides the same number of calories, the virtual water footprint of a vegetarian diet is considerably smaller—2,300 to 2,700 liters (610–715 gallons) per day, or enough water to fill 8 or 9 bathtubs. The difference is due to the crops consumed by the animals in producing the meat.

To produce **1 small** pizza margherita requires

1,260 liters (333 gallons)
of water

Water Breakdown per Ingredient

Mozzarella Cheese	Wheat Flour	Tomato Purée
50%	**44%**	**6%**

The water footprint of a pizza depends largely on its ingredients and toppings. If we take the simplest of pizzas, the margherita—made with a wheat flour crust, tomatoes and mozzarella—the biggest proportion of virtual water lies in cultivation of the tomatoes and the feed crops of the dairy cows that provided the milk for the cheese. The global average water footprint of one 725-gram (26-ounce) pizza margherita is 1,260 liters (333 gallons). The mozzarella represents about 50% of the total water used, the wheat flour 44% and the tomato purée about 6%.

To produce **1 cheeseburger** requires

**2,400 liters (634 gallons)
of water**

150 grams (5 ounces)
Ground Beef
2,300 liters (634 gallons)

10 grams (0.35 ounces)
Slice of Cheese
50 liters (13 gallons)

Bun
50 liters (13 gallons)

A typical 160-gram (5-ounce) cheeseburger requires 2,400 liters (634 gallons) of water to produce—many times the amount of water the average North American uses every day for drinking, bathing, washing dishes and flushing the toilet. Most of the water is needed for producing the beef, which has a water footprint of 15,415 liters (4,072 gallons) for every 1-kilogram (2-pound) portion. Including a slice of cheese and a bun adds about 100 liters (26 gallons) of virtual water. Toppings such as tomato, lettuce, pickle and mayo further add to the water footprint.

To produce **1 egg (60 grams/2 ounces)** requires

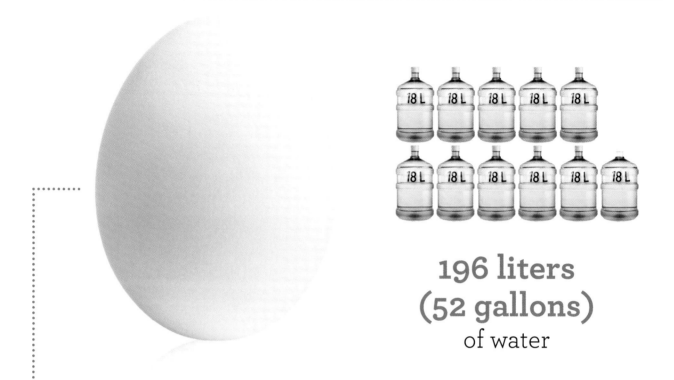

**196 liters
(52 gallons)**
of water

Since 1996 the global water footprint for layer chickens is

7%
of the total
footprint of all
farm animals

On average worldwide, one 60-gram (2-ounce) egg requires 196 liters (52 gallons) of water to produce. Most of this is required for feeding the chickens. Chickens are a thirsty bunch—in the 10 years following 1996, the water footprint of laying chickens was about 7% of the total water footprint of all the farm animals in the world.

To produce **1 stick (250 grams/9 ounces)** of butter requires

1,387 liters (366 gallons) of water

The global average water footprint of whole cow's milk is about 940 liters (248 gallons) for every kilogram (2 pounds). Butter derived from whole milk accounts for about 28% of this amount, and the remaining 72% goes to skim milk. A 250-gram (9-ounce) stick of butter takes about 1,387 liters (366 gallons) of virtual water to produce.

To produce **1 kilogram (2 pounds)** of dry pasta requires

1,850 liters (490 gallons) of water

250 L · 250 L · 250 L · 250 L · 250 L · 250 L · 250 L · 250 L

·········· *Which is* ··········

925 liters (245 gallons) per package (500 grams/ 1 pound)

or

1.85 liters (½ gallon) for a single spaghetti noodle

At its most basic, pasta is made from durum wheat semolina and water. The global average water footprint of wheat is 925 liters (244 gallons) per 500-gram (18-ounce) package, which is enough pasta to serve 4 people. The water footprint of pasta will vary depending on the sauce you serve with it. A simple tomato sauce includes the water footprint of tomatoes, and pasta primavera that of fresh vegetables. Sauces such as Alfredo and bolognese add the water footprint of cream and beef, respectively.

To produce **1 kilogram (2 pounds)** of milled rice requires

250 L · 250 L · 250 L · 250 L · 250 L
250 L · 250 L · 250 L · 250 L · 250 L

2,500 liters (660 gallons) of water

Millet
**4,478 liters/kilogram
(1,183 gallons/2 pounds)**

Sorghum
**3,048 liters/kilogram
(805 gallons/2 pounds)**

Rolled Oats
**2,416 liters/kilogram
(638 gallons/2 pounds)**

Rye
**1,930 liters/kilogram
(510 gallons/2 pounds)**

Wheat Flour
**1,849 liters/kilogram
(488 gallons/2 pounds)**

Barley
**1,977 liters/kilogram
(522 gallons/2 pounds)**

A staple food for 3 billion people, rice is also one of the largest water consumers in the world. On average, rice requires 2,500 liters (660 gallons) of water to yield only 1 kilogram (2 pounds) of milled rice.

To produce **1 kilogram (2 pounds)** of pet food requires

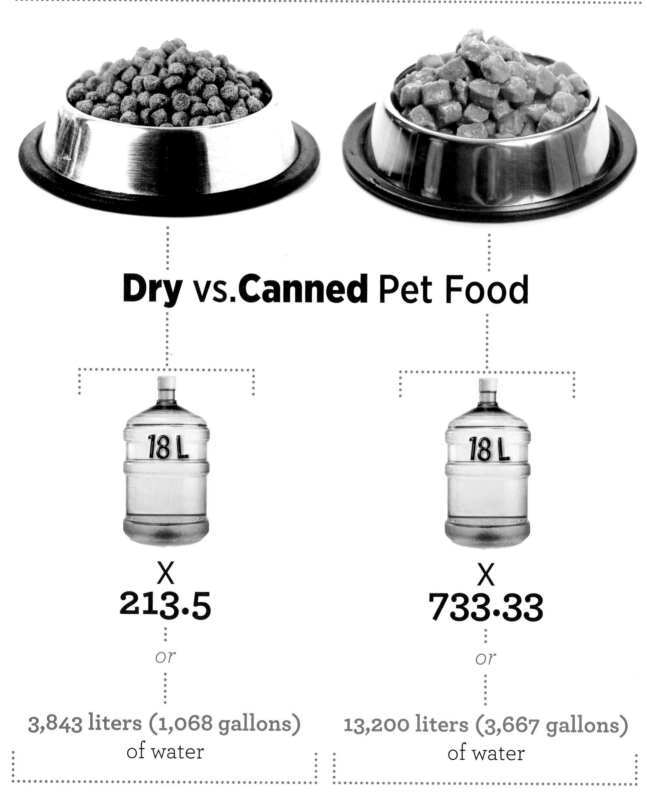

Dry vs. **Canned** Pet Food

18 L

X
213.5

or

3,843 liters (1,068 gallons) of water

18 L

X
733.33

or

13,200 liters (3,667 gallons) of water

Dry kibble for cats and dogs contains far less moisture (only 6–10%) than the moist canned variety (60–90%). The water footprints for dry and canned pet foods are comparable to those for cereals and beef. The pet food option with the lowest water footprint is dry vegetarian kibble.

To produce a **1-liter (34-fluid-ounce)** bottle of olive oil requires

15,100 liters (4,000 gallons)
of water

The global average water footprint of olives is 3,020 liters (800 gallons) of water per kilogram (2 pounds). Since it takes roughly 1 kilogram of olives to produce 200 grams (7 ounces) of oil, a 1-liter (34-fluid-ounce) bottle requires about 15,100 liters (4,000 gallons) of water.

To produce **1 kilogram (2 pounds)** of sugar from **sugar beets** requires

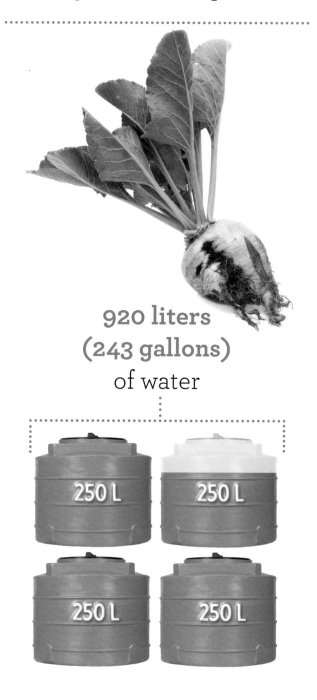

920 liters (243 gallons) of water

To produce **1 kilogram (2 pounds)** of sugar from **sugar cane** requires

1,800 liters (475 gallons) of water

It takes about 210 liters (55.5 gallons) of water to produce 100 grams (3.5 ounces) of refined sugar. Sugar cane is often grown using irrigation and so has a higher blue water footprint than sugar made from

sugar beets, which are grown in temperate climates that receive more rain and consequently have a lower water footprint.

To produce 1 chocolate bar (200 grams) requires

1,700 liters (449 gallons) of water

Chocolate is made from ingredients with high water footprints: cocoa paste, cocoa butter and cane sugar.

Cocoa beans are native to rainforests and require vast amounts of water to thrive.

To produce **one 500-milliliter (17-fluid-ounce) bottle** of cola requires

175 liters (46 gallons) of water

Actual Water
500 milliliters (17 fluid ounces)

Manufacturing & Supply Chain
11.5 liters (3 gallons)

Production of flavoring
163 liters (42 gallons)

Cola is almost entirely water, so a half-liter (17-fluid-ounce) bottle effectively contains a half-liter of water. That's the direct water input. But cola is not just water in a bottle. When you include the production of all of the flavoring ingredients (the highest consumptive factor here), the manufacturing and the supply chain, each bottle requires about 175 liters (46 gallons).

To produce **one 750-milliliter bottle** of whisky requires

1,218 liters (322 gallons) of water

········ *Other alcoholic beverages require per liter* ········

| 770 liters (203 gallons) of water for **Wine** | 370 liters (98 gallons) of water for **Beer** | 302 liters (80 gallons) of water for **Vodka** | 433 liters (114 gallons) of water for **Gin** |

Water is used in all aspects of making whisky. It takes about 960 liters (254 gallons) just to grow the barley to make a 750-milliliter (25.4-fluid-ounce) bottle of single malt. Water used to soak and germinate the barley, to distill, mature and bring the whisky to bottling strength—40% alcohol by volume—adds up to a virtual water total of 1,218 liters (322 gallons). The Scotch whisky industry uses 61 billion liters (16 billion gallons) of water annually.

"Filthy water cannot
be washed."

—West African proverb

Manufacturing and Farming

Everything we make, use and eat requires water. When we talk about energy we rarely consider the amount of water required to produce it, whether it is gas for the car or electricity for our homes. The paper on which these words are printed required water for its production—an enormous amount, in fact—not counting the water required to grow the trees from which the paper was made. Cars, our most popular form of transportation, require a lot of water just to fabricate the steel, rubber and plastic parts that go into them. The gas on which cars run uses water in the manufacturing process, and yet more water if the gas is steamed out of tar sands or made from corn that is cultivated and grown to be used as an additive to refined gasoline. Whereas the water used in making fuel is contaminated forever, water used to turn the turbines in a hydroelectric power plant is "borrowed" water, returned back to the environment after it is used. Groundwater aquifers from which the irrigation of corn for fuel may be drawn is gone (for at least our lifetime) unless it is replenished naturally at the same rate it is withdrawn.

Surface vs. In Situ Mining of Oil Sands

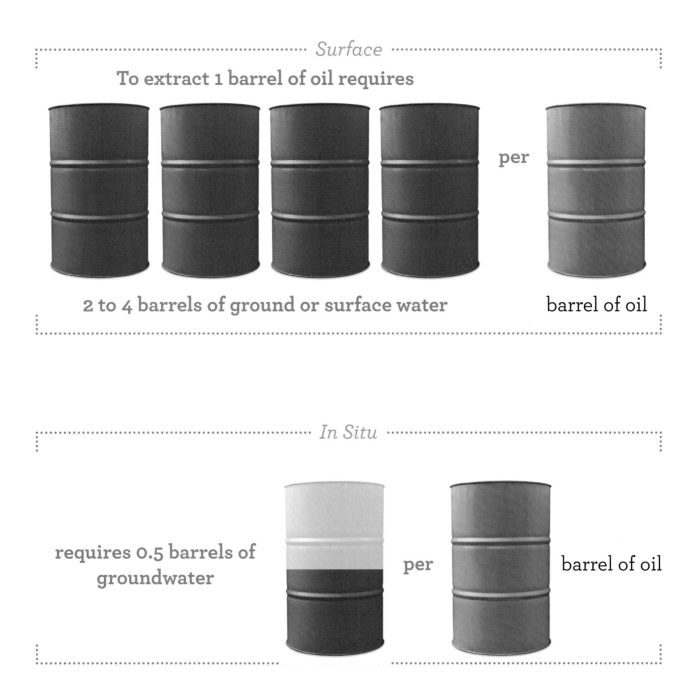

Surface

To extract 1 barrel of oil requires

per

2 to 4 barrels of ground or surface water

barrel of oil

In Situ

requires 0.5 barrels of groundwater

per

barrel of oil

Bitumen, or heavy crude oil, can be extracted from oil sands using either surface or in situ mining. Surface mining is used when the crude oil is close to the surface; it requires the net equivalent of 2 to 4 barrels of water per barrel of oil produced to separate the bitumen (10%) from the sands (83%). In situ operations involve drilling wells to reach deeper oil sand deposits and injecting steam to liquefy the oil before pumping it to the surface. This method requires only 0.5 barrels of water per barrel of bitumen produced. Most of this is groundwater, and much of this water is reused before being disposed of deep underground. It takes about 2 tonnes (2.2 tons) of oil sand to produce a barrel of oil.

To produce **1 liter** of soybean-based biodiesel fuel,

It requires 11,397 liters (3,010 gallons) of water

If stacked end-to-end, that would be enough 1-liter bottles to span the height of the Empire State Building 11 times!

Biofuels made from crops such as soybeans, corn and sugar are placing increasing pressure on agricultural land. While renewable, biofuels have a much larger water footprint than conventional fossil fuels. Based on the energy it creates, soybean-based biodiesel has a water footprint more than 3,000 times that of crude oil.

Biofuel use by 2030 is projected to be

3.2 million barrels per day

5%

This will be the proportion of vehicles run on biofuels worldwide.

Production of these fuels could consume 20% of all water now used for agriculture worldwide

The International Energy Association predicts that 5% of cars and trucks worldwide will be run on biofuels by 2030. This amounts to a staggering 3.2 million barrels of fuel per day. If production processes and technologies do not advance, this means production of biofuels could consume 20%

of all the water now used for agriculture worldwide. The U.S. Energy Independence and Security Act of 2007 mandated annual production of 57 billion liters (15 billion gallons) of ethanol from corn by 2015, with an additional 60.6 billion liters (160 billion gallons) of biofuels by 2022.

To produce **4 new rubber tires** requires **7,850 liters (2,074 gallons)** of water

The global average water footprint of industrial products is 80 liters (21 gallons) of water used for every dollar of purchase price. It takes 459 liters (121 gallons) of virtual water to make 1 kilogram (2 pounds) of synthetic rubber, and the total for making and fitting your car with four new tires can top 7,850 liters (2,074 gallons).

69,033 liters (18,237 gallons)

Amount of water used by an **average car** in its lifetime

Automobile Life Cycle Water Consumption

60,000 liters (15,850 gallons)
Use (based on 160,000 km), primarily for fuel production

259 liters (68 gallons)
End of Life

5,570 liters (1,471 gallons)
Material Production

894 liters (236 gallons)
Production of Parts

2,310 liters (610 gallons)
Assembly of Vehicle

An analysis of a typical car's water footprint, including water used in parts production, vehicle assembly, vehicle use, and disposal and recycling at the end of its life, found that the greatest water consumption (87%) occurs during the use phase, when consumers are driving. This is largely because of the amounts of water required for fuel production.

5,000,000+

The estimated number of people who will have been relocated by 2020, by construction of the Three Gorges Dam, the largest dam in the world.

The reservoir of the dam **has flooded**

13 cities

140 towns

1,350 villages

and **eradicated** 1,300 archaeological and cultural sites

Part of a $50 billion project to divert up to 4.5 trillion liters (1.2 trillion gallons) of the Yangtze River annually to parched areas of the north, the Three Gorges Dam is an engineering marvel capable of generating 22,500 mW of electricity. But its 660-kilometer (410-mile) reservoir, which started filling in 2003, has flooded 13 cities, 140 towns and 1,350 villages, eradicated 1,300 archaeological and cultural sites, increased the risk of landslides, and wrought ecological havoc on several endangered species. It will eventually displace several million people.

To produce **1 kilogram (2 pounds)** of paper requires

3,000 liters (793 gallons) of water

The pulp and paper industry **uses more water to produce a tonne of product** than any other industry.

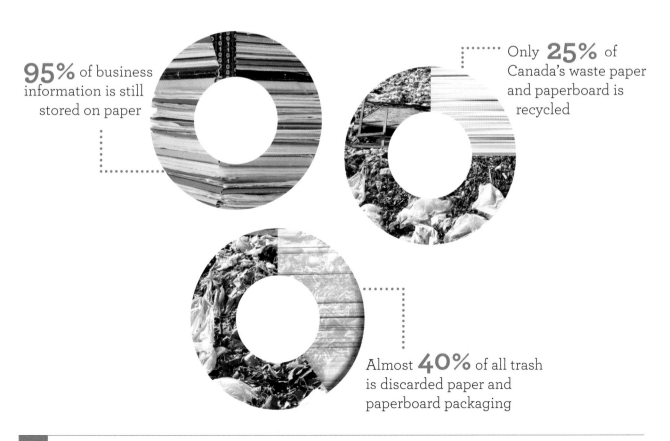

95% of business information is still stored on paper

Only **25%** of Canada's waste paper and paperboard is recycled

Almost **40%** of all trash is discarded paper and paperboard packaging

5.4 million tonnes (3 tons)
of paper and paperboard are produced annually

48 kilograms (106 pounds)
Average global per capita paper use

Every year in the United States, more than

359 million magazines
2 billion books &
24 billion newspapers
are published

The average Internet user prints

28 pages daily

A typical American lawyer uses

900 kilograms (1,984 pounds)
of paper every year

Next to electricity generation, the pulp and paper industry is the single largest consumer of water in industrialized countries. The effluents and toxins released rank the industry as one of the most notorious polluters in the world. A staggering amount of water is used in nearly every step of manufacturing pulp and paper—an average of 3,000 liters per kilogram (793 gallons per 2 pounds) of finished paper.

Seismologists have now determined that the Great Sichuan Earthquake of 2008 **may have been triggered** by the weight of **320 million tonnes (352 million tons) of water** behind the Zipingpu Dam

7.9

Magnitude of the Great Sichuan Earthquake of 2008 — the 21st deadliest earthquake of all time.

On May 12, 2008, a devastating earthquake struck Sichuan province in China, killing an estimated 70,000 people, leaving almost 5 million homeless, and causing an estimated $20 billion in property damage. Seismologists calculated that the quake may have been triggered by the weight of 320 million tonnes (352 million tons) of water behind the Zipingpu Dam, creating pressures along a fault line that were 25 times those of normal tectonic forces. The 156 meter (515 feet) tall dam was built on the Min River only 500 meters (550 yards) from the fault line. The Chinese government has denied any causal link between the dam and the quake.

43% of water used globally for irrigation is groundwater

Countries That Draw the Most Groundwater for Irrigation

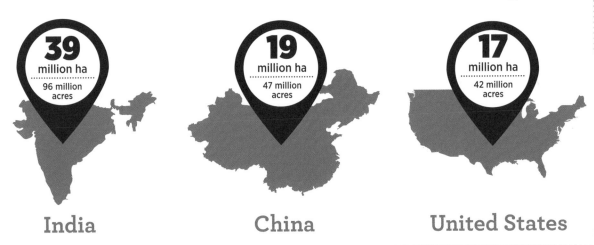

39 million ha
96 million acres

India

19 million ha
47 million acres

China

17 million ha
42 million acres

United States

One-third of the world's population depends on groundwater for survival, and by far the most common use of groundwater is for irrigation. The countries that draw the most groundwater for irrigation are India, at 39 million hectares (96 million acres); China, at 19 million hectares (47 million acres); and the United States, at 17 million hectares (42 million acres). Groundwater use is increasing as a percentage of total irrigation, which is leading to depletion of aquifers that took thousands of years to fill.

70% of global freshwater withdrawal is for irrigation

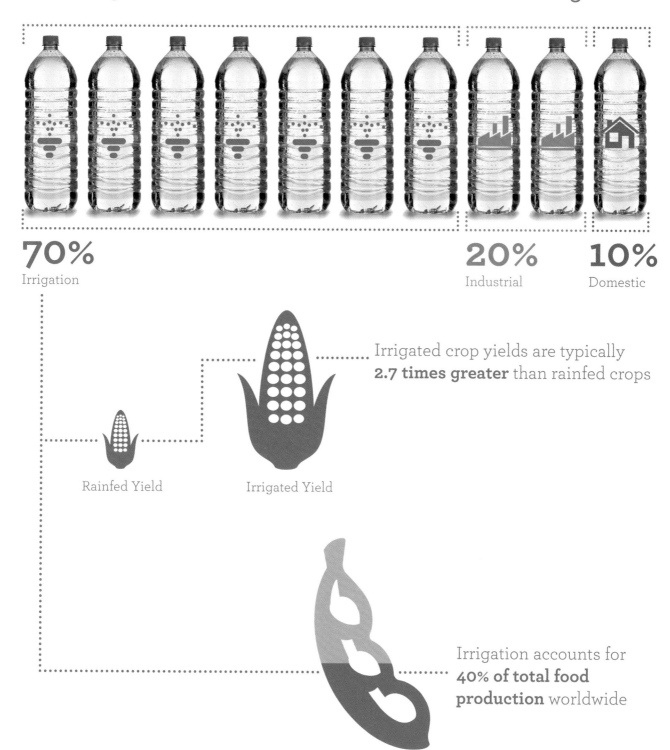

70%
Irrigation

20%
Industrial

10%
Domestic

Rainfed Yield

Irrigated Yield

Irrigated crop yields are typically **2.7 times greater** than rainfed crops

Irrigation accounts for **40% of total food production** worldwide

Global freshwater consumption is overwhelmingly linked to growing food, with water for crop irrigation accounting for 70% of the blue water withdrawn annually. Irrigated crop yields are typically 2.7 times greater than rainfed crops. While irrigation accounts for only 20% of agriculture's annual total blue and green (rain) water consumption, it contributes 40% of the total food produced worldwide.

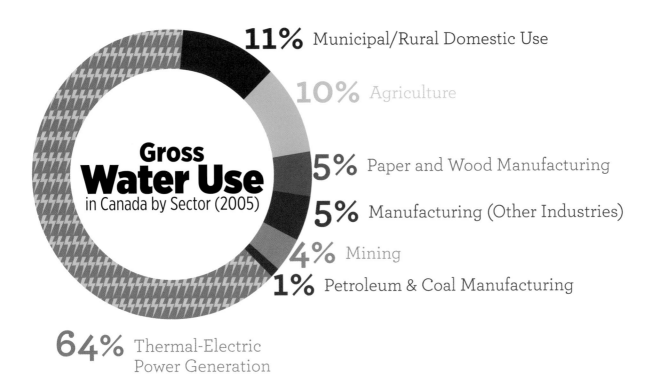

Gross Water Use in Canada by Sector (2005)

11% Municipal/Rural Domestic Use

10% Agriculture

5% Paper and Wood Manufacturing

5% Manufacturing (Other Industries)

4% Mining

1% Petroleum & Coal Manufacturing

64% Thermal-Electric Power Generation

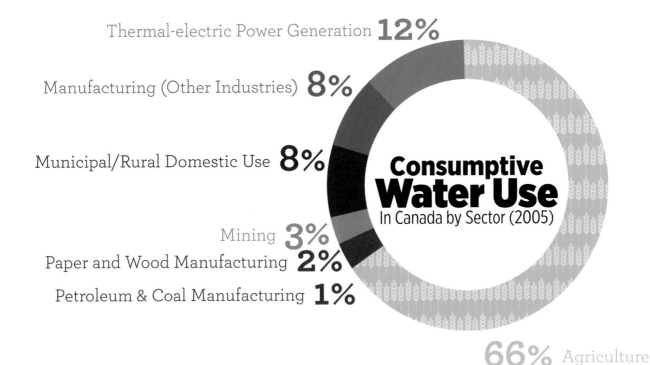

Consumptive Water Use In Canada by Sector (2005)

Thermal-electric Power Generation 12%

Manufacturing (Other Industries) 8%

Municipal/Rural Domestic Use 8%

Mining 3%

Paper and Wood Manufacturing 2%

Petroleum & Coal Manufacturing 1%

66% Agriculture

Gross water use refers to water that is returned to its source after being used. *Consumptive water use* refers to water that is effectively withdrawn from the source and not returned. Thermal-electric, nuclear and fossil fuel power-generating facilities withdraw massive amounts of water for cooling. Agricultural use of water is overwhelmingly related to crop irrigation and raising livestock.

40% of industrial water consumption in the United States is used to cool power plants

In industrialized nations, industry tends to consume more than half of the water available for human use, every year.

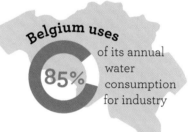

Belgium uses **85%** of its annual water consumption for industry

5 countries account for more than half of the world's generated hydropower

Canada

Brazil

China

United States

Russia

Cooling power plants represents 40% of all industrial water use in the United States. Most of the hundreds of millions of liters of water used every minute of every day to generate electricity is returned to its source. The rest evaporates, and this is especially true for hydroelectric power. Five countries—China, Canada, the United States, Brazil and Russia—account for more than half of the world's generated hydropower.

760
liters
(200 gallons)

The **daily water consumption** for an average hotel room

A 250-room hotel will **heat 40,000 liters (10,600 gallons) of water** every day, just to do the laundry

Which is roughly the tank capacity of an 18-wheeler truck

The average hotel room accounts for around 760 liters (200 gallons) of water daily. While the greatest use of water in a hotel is showering (56%), an average 250-room hotel will heat about 40,000 liters (10,600 gallons) of water daily—roughly the capacity of an 18-wheeler tanker truck—just to do the laundry. Daily water consumption for hotel laundry in the U.S. is estimated to be 4.5 billion liters (1.2 billion gallons).

2,500

Number of **public drinking fountains—called** *nasones*— in Rome

First built in **1874**

Peschiera Reservoir

Rome

280 are in Rome's historical center

The water travels 110 kilometers (68 miles) before emerging from a *nasone*

Stout cast-iron tubes standing about a meter (3 feet) tall and dispensing clean, chilled freshwater, Rome's drinking fountains are an engineering marvel. Dubbed *nasone* ("big nose") for the shape of their spout, the first 20 were built in 1874, and they have since become a symbol of the city. Today about 2,500 of these fountains form one of the most extensive networks of free drinking water in the world. The water for the *nasoni* comes from springs in Peschiera, approximately 110 kilometers (68 miles) from Rome.

Murray–Darling Basin

The size of France + Spain combined

Responsible for

40% of Australia's annual agricultural output

In 2006 these farms used

83% **of the basin's total freshwater withdrawal**

The Murray–Darling Basin is the catchment area for Australia's two biggest rivers, the 2,740-kilometer (1,700-mile) Murray and the 2,520-kilometer (1,566-mile) Darling. The basin is the size of France and Spain combined and produces 40% of Australia's annual agricultural output, with livestock, dairy, cotton and rice farming being the major activities. As of 2006 there were 61,033 farms operating in the basin, using 83% of the basin's total freshwater withdrawal.

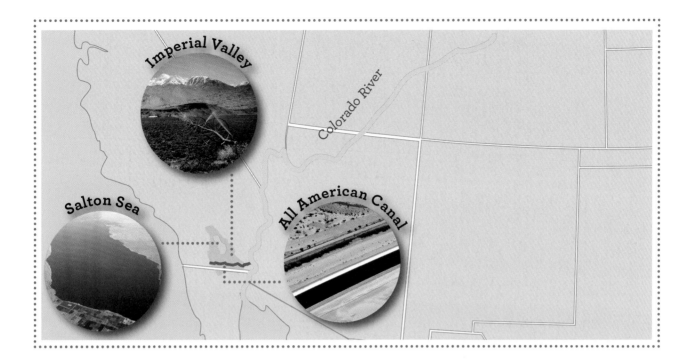

3.7 cubic kilometers (1 cubic mile)

Amount of Colorado River water used to irrigate California's Imperial Valley every year

The Imperial Valley produces:

Annual crop worth
over $1 billion

80% of the United States winter fruit and vegetables

An oasis of fertile farmland surrounded by desert, the Imperial Valley was created in the 1930s when the massive Imperial Dam was built on the border of California and Arizona. Today the valley is one of the most productive farming regions in the United States, with an annual crop output worth over $1 billion. Farms in the valley get nearly all their water from the Colorado River. As much as 3.7 cubic kilometers (almost 1 cubic mile) of Colorado River water is used every year to irrigate farmland in the valley.

77.8%

Kansas's dependency
on groundwater

Groundwater as percent of total water withdrawal

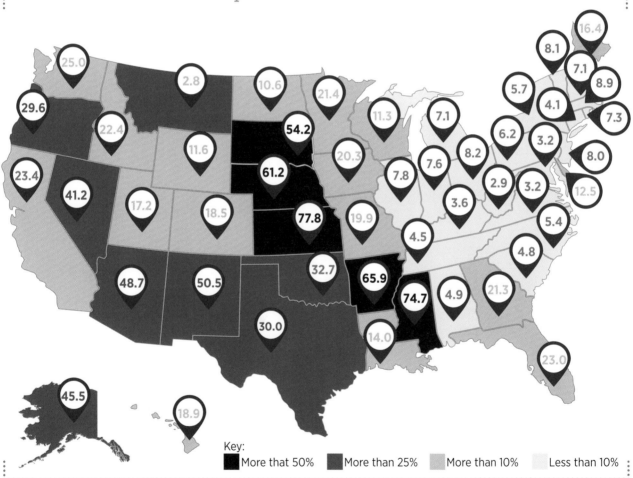

Key:
■ More that 50%   ■ More than 25%   ■ More than 10%   □ Less than 10%

The drier desert states of the U.S. Southwest draw half of the water needed for agriculture, industry and domestic use from groundwater sources. With its semi-arid climate, limited surface water supplies and intensive wheat farming, Kansas relies on groundwater from the shrinking Ogallala Aquifer for more than three-quarters of its water needs. Five states—California, Texas, Nebraska, Arkansas and Idaho—account for nearly half (48.2%) of daily U.S. groundwater withdrawal.

CONCLUSION
Water and the Future

Agriculture accounts for 75% to 85% of all the water used by humanity. How much water is that, exactly? Imagine a canal 10 meters (11 yards) deep and 100 meters (110 yards) wide. That's a very big canal, considering that the Panama Canal is less than 35 meters (40 yards) wide. To hold the amount of water used each year to grow our food, our imaginary canal would have to be 8 million kilometers (5 million miles) long. It would circle the planet 192 times.

To meet the future food needs of 2 or 3 billion additional people by the end of this century, our water-for-agriculture canal will have to be extended another 4 or 5 million kilometers (2.5 to 3 million miles). There simply isn't enough freshwater available to fill a 12 to 13 million–kilometer (6.5 to 7 million–mile) canal. We're already struggling to fill the one we have now.

The success and prosperity of many parts of the world are directly linked to overdrawing of their water resources. This can't continue. The U.S. Southwest, the Mediterranean region, countries in the Middle East and southern Africa, parts of China and India, among other regions, are living in a "water bubble"—their current rates of water usage are unsustainable. Over the past 50 years cheap energy and technology have made it possible to move huge amounts of water without regard to natural limits or the sustainable yield of water sources. These water bubbles are beginning to burst, pushed by increasing temperatures and changes in precipitation patterns resulting from climate change.

The planet's global average temperature has risen 0.8°C (1.4°F) over the past century. This increase is due mainly to the burning of fossil fuels, which emits large volumes of carbon dioxide (CO_2). Measurements show that there is now 42% more CO_2 in the atmosphere than there was in 1900, and for millions of years before that. CO_2 is the main greenhouse gas that traps the sun's heat. Thus with so much rapidly added CO_2 in the atmosphere to trap the sun's heat, it's inevitable that this increase will have a substantial impact on the climate. One easily measured impact is warmer temperatures.

Canada is now 1.6°C (2.9°F) warmer, and its northern regions are 2.5°C to 3°C (4.5°F to 5.4°F) hotter. The United States is on average 1°C (1.8°F) warmer, with Alaska even more so. We have already begun to experience the impact, including increases in extreme temperatures, especially heat waves, and changes in precipitation patterns, including more and longer droughts and more and bigger floods.

BURSTING WATER BUBBLES AND THE NEW NORMAL

In 2012 the United States suffered its worst drought in 50 years, with nearly two-thirds of the country facing water shortages. Rationing and water-use bans were in place in many parts of the country. Operations at U.S. power plants and drilling for oil and gas had to be scaled back because of water restrictions.

The recent droughts in California, the Southwest and the Midwest are part of a larger drought that has lasted roughly 13 years. Over the past 10 years, irrigation has been discontinued for at least 2 million acres of farmland in the United States because of water shortages. Even the Great Lakes, with 20% of the world's freshwater, have been at record low water levels for the past 14 years. This is the new normal, resulting from rising temperatures caused by climate change. Climatologists now calculate that most of the western U.S. and southern Canada will face drier conditions in coming years. Historically these regions are naturally dry; over the past 2,000 years, mega-droughts lasting decades were common. Climate change is simply making this natural dryness worse.

In 2013 California experienced its worst drought in history. Nearly 25 million people were told there wasn't enough water, even after water allocations to agriculture were cut by 50%. By early 2014, the region was in a state of drought emergency. There was barely any snow in the Sierra Nevada Mountains—the "water towers" of California—which normally hold large stores of winter snow and ice that become vitally important freshwater in spring and summer.

Hardest hit by the 2013–14 drought was California's $44.7-billion (2012) agriculture industry, which supplies nearly half of all U.S. fruits, veggies and nuts. California farms use about 80% of the state's water, which is moved from its natural sources to other areas via a complex system of rivers, canals and reservoirs. Water is heavy, so moving large volumes long distances takes a great deal of energy. It's no surprise that 20% of the state's entire energy use goes to moving water.

During dry years, and especially during this record drought, farmers pumped huge quantities of groundwater. Urban areas and industry also use groundwater; in dry years the state meets 60% of its water needs this way. There is no cost for groundwater and no real restrictions in California on how much can be used. Not surprisingly, the state's groundwater is being depleted far faster than natural recharge rates. Large areas are sinking or subsiding as the water is drained from the ground. Scientists estimate that without drastic reductions in withdrawals, California is just two to three decades away from total groundwater depletion.

Many western and Midwestern states are also living in water bubbles by taking more water than is sustainable. Water levels are falling in the Ogallala Aquifer, which lies under eight states from South Dakota in the north to Texas in the south—an area of nearly half a million square kilometers (174,000 square miles). It is one the world's great underground reservoirs of freshwater, filled mainly with meltwater from the thick ice sheets that covered North America 12,000 years ago. The Ogallala supplies 30% of all irrigation water in the United States.

Since 1960 water levels in the Ogallala have been falling, and about one-third is now gone. At current rates the aquifer will be mostly empty in less than 50 years, which will have a devastating impact on U.S. food production. Water experts recommend an immediate 20% reduction in the amount of water taken in order to prolong its life. And the United States is not alone in this dilemma. At least 17 other countries, including China and India—which represent more than half of the world's population—are living in water bubbles where food is being produced by depleting their water reserves.

With 38 million people and increasing water scarcity, California will likely be the first U.S. state forced to come to grips with its collapsing water bubble. Many California farmers say they will not survive another year of drought. More than 90% of California's water footprint (consumption) is associated with agricultural products. The meat and dairy industries account for nearly half of the state's water footprint because of the water-intensive feed required to raise the animals.

Direct household water consumption represents 4% of the state's water footprint, and much of this is for watering lawns and gardens. Home water use varies tremendously within the state. In Sacramento, where half the homes still don't have water meters, residents use 1,000 liters (280 gallons) a day per person, while residents in San Francisco use only 400 liters (100 gallons). Meanwhile, Palm Springs, with its acres of green lawns and lush golf courses in the desert, consumes a staggering 2,780 liters (740 gallons) a day per person, five times more than residents of San Jose and Los Angeles.

Across North America lawns and gardens account for half the water used in the average home, and much more in the dry U.S. Southwest. Covering 16.6 million hectares (41 million acres), lawns are America's largest irrigated "crop." Lawns are also a major source of water pollution from overuse of fertilizers and pesticides. The Association of California Water Agencies estimates that a small lawn of 90 square meters (1,000 square feet) requires about 132,500 liters (35,000 gallons) a year in watering, over and above the rain it receives.

THE GLOBAL TRADE IN WATER

The United States and Canada are major exporters of wheat, corn and soybeans, as well as meat products and other foods. All of these require large amounts of water to grow, which means that food exports are, effectively, water exports. These virtual water exports are estimated to equal more than twice the annual flow of the Mississippi River. Since virtual water is embedded in nearly everything, mighty rivers of virtual water—made up of grains, furniture, electronic gadgets and more—are flowing from one country to another all around the planet.

In 2013 China overtook the United States as the world's largest exporter of food and goods. However, the U.S. remains the biggest importer, buying $472 billion more products and services than it exports. This means that America actually imports more virtual water than it exports, as do many European countries.

China is known as "the world's factory" because nearly everything now seems to be made there. While that has resulted in short term expansion for the country's economy, it has been very damaging to its environment, and especially its river systems. Due to unsustainable withdrawals for industry and agriculture, China has fully depleted over 27,000 rivers since the 1950s. And the remaining 23,000 rivers are not in great shape. Some experts predict that by 2025 not a single Chinese river will reach the sea, except during floods, with tremendous effects on the coastal fisheries there.

Surprisingly, Australia—the driest continent on the planet—is the world's biggest net exporter of virtual water, in the form of wheat, cotton, coal and other products. Canada is the second largest net exporter, mainly as food and energy. Alberta, one of Canada's driest provinces, exports nearly 17 billion cubic meters (600.4 billion cubic feet) of virtual water every year in the form of meat, grain and oil, of which America purchases the lion's share.

At one level, global trade is simply a trade of money for water and water for money. A number of countries, particularly those in the Middle East, must trade money for water because they don't have enough water to grow their own food. These include Israel, Libya, Kuwait, Qatar, Jordan, Egypt and Saudi Arabia. Saudi Arabia once produced most of its own food and even exported crops such as wheat. Deep beneath the desert sands existed an ancient series of aquifers large enough to hold all the water in Lake Erie. However, most of the water is now gone after 40 years of pumping out water with very little going back in, since it hardly ever rains there.

Saudi Arabia has given up on its goal of being self-sufficient in food production; in 2013 it decided to stop growing wheat. The country is relying increasingly on food imports, and to secure those imports it is spending tens of millions of dollars to buy vast tracts of farmland in Africa, Pakistan and the Philippines. Saudi Arabia has plenty of land—it's really buying water. Other Middle Eastern countries, as well as China, India and investors from the United States, have acquired an estimated 45 to 50 million hectares (111 to 123.5 million acres) of farmland in the past few years, in Africa, Latin America and southern Asia.

Like most countries, Egypt can't make or find more water within its borders. It could do a better job of managing water. It is the world's second largest exporter of oranges, shipping 998,000 tonnes (1.1 million tons) in 2013. Since 1 tonne (1.1 tons) of oranges has a water footprint of 500,000 liters (132,100 gallons), by selling these oranges Egypt is selling huge amounts of water to other countries. The biggest reason why dry countries sell water-rich food and products to wet countries is because water is egregiously undervalued. In many places, including the United States and Canada, it's either free or ridiculously underpriced. Bottled-water companies in Canada pay just $3.71 for 1 million liters (264,200 gallons). In drought-stricken California, anyone can pump up all the groundwater they want—no charge, no limits, no regulations to speak of.

Nearly every country subsidizes the costs of water and energy. In many places agriculture and industry can help themselves to unsustainable amounts of water and governments will even pay most of their energy costs for taking it. The final buyer is the beneficiary of these subsidies, in the form of low-cost products. Yet the actual costs, including the impacts of depleted or polluted waterways, are not passed on to the final purchaser. The $6 price tag on a pair of jeans imported from Bangladesh does not include the cost of water pollution and falling water tables.

The global trade system ignores these costs and their impacts, resulting in unsustainable water use throughout the world. And so we end up with poor dry countries in Asia and Africa selling water-rich products to wealthy wet countries such as Canada and the United States. However, international transfers of water in the form of food and other products have become a necessity. Jordan imports 6 times more virtual water than it could sustainably pump within its own boundaries. The solution is not to stop the virtual water trade but to turn it into water-smart trade. That means valuing and managing water appropriately.

Most of the serious conflicts over water are local or regional—and often the result of mismanagement. Syria is an example. Mismanagement of water is one of the causes of the Syrian civil war. Syria, along with the entire Middle East and the Mediterranean region, has experienced significantly drier winters over the past 40 years. This is almost certainly the result of global warming, which tends to make dry regions drier and wet regions wetter. Between 2010 and 2012 alone, water scarcity sparked conflicts in India, Pakistan, Brazil, Afghanistan, Sudan and a dozen other countries.

Climate change will make it more difficult to grow food in many parts of the world. Add to this growing populations and the number of local conflicts is sure to increase. Preventing this grim future will be possible only through cooperation and responsible water management, including ensuring enough water for nature. In the past few years, common concerns over water have brought together Palestinians, Egyptians, Jordanians and Israelis to protect and restore the Jordan River valley. Unfortunately this did not lead to progress on the political front in the Israel–Gaza land and water conflict.

Leaving enough water for nature is essential for our own health and prosperity. Economic theory views the environment as a subset of the economy. Political and business decisions are made on that basis every day, when the reality is exactly opposite: the economic system we've created is wholly dependent on the natural environment. As a direct result, ecologically ignorant policies are responsible for climate change, desertification, the extinction crisis, water and ocean pollution and the destruction of forests. The human population of our planet cannot be sustained while its basic life-support system is being continuously unraveled.

Water underpins every aspect of our societies. No one should be in a position of power and influence—this includes politicians, public officials and business leaders—without a high level of literacy regarding water and ecology issues. Such literacy should be part of everyone's basic education.

Water has been held as sacred for most of humanity's history for good reason. It's definitely time for an attitude adjustment.

Water-saving Tips

Everything we do, from eating and drinking to our daily commute, uses water (a lot goes into producing gas and energy). The clothes we wear require water to produce, and the products we buy also use water in their manufacture. Our daily water footprint includes more than the water we drink and use in our homes. Besides choosing less water-intensive products (for example, polyester versus cotton), we can take steps to limit the size of our water footprint in everyday activities such as lawn watering, clothes washing and flushing the toilet.

By knowing how dependent we are on water, not only for our health but for our modern lifestyles, we can change what we do. We can reduce wastage, change habits and make water-smart product purchases, all of which can save both water and money.

Bathroom

- Take shorter showers and limit baths.

- Don't let the water run while brushing your teeth.

- Install aerators on faucets and use water-saving showerheads.

- Replace older toilets with newer water-saving ones that use only 6 liters (1.6 gallons) per flush. The initial expense is worth it in the long run.

- Avoid flushing the toilet unnecessarily. Dispose of tissues, insects and other such waste in the trash rather than the toilet.

- Check for toilet leaks by adding food coloring to the tank. If the toilet is leaking, color will appear in the bowl within 30 minutes. (Flush as soon as test is done, since the food coloring may stain.) Check for worn-out, corroded or bent parts. Most replacement parts are inexpensive, readily available and easily installed.

- Make sure that your home's pipes are leak-free. Many homes have hidden water leaks. Check your water meter and then read it after a two-hour period when no water is being used. If the reading is not exactly the same, there is a leak.

- Repair dripping faucets by replacing washers. One drop per second can waste 10,225 liters (2,700 gallons) per year, which will add to the cost of water and sewer utilities and/or strain your septic system.

- Use the minimum amount of water for a bath by filling the tub only one-third full. Stopper the tub before turning on the water.

Kitchen and Laundry

- Dishwashers should be fully loaded for optimum water conservation. There is usually no need to pre-rinse dishes.

- When washing dishes by hand, don't leave the water running for rinsing. If you have a double sink, fill one side with soapy water and one with rinse water. If you have a single basin, place the washed dishes in a rack and rinse them with a spray device or a pan full of hot water.

- Compost organic waste instead of using kitchen-sink garbage disposal units, which require lots of water to operate properly and put a strain on purification and septic systems.

- Don't let the faucet run while you clean vegetables. Rinse them in a stoppered sink or a pan of clean water.

- Keep a bottle or pitcher of drinking water in the fridge. Running tap water to cool it off for drinking is wasteful. Take refillable water bottles with you on outings.

- When washing clothes, avoid the permanent press cycle, which uses an additional 20 liters (5 gallons) for the extra rinse. For partial loads, adjust water levels to match the size of the load. Replace old laundry appliances. New Energy Star–rated washers use 35% to 50% less water and 50% less energy per load. If you're in the market for a new clothes washer, consider buying a water-saving front-loading machine.

Outdoors

- Attach a rain barrel to downspouts to collect runoff from gutters (eavestroughs). Use the collected water for the garden.

- Plant drought-resistant lawns, shrubs and plants. If you are planting a new lawn or reseeding an existing lawn, use drought-resistant grasses.

- Plant slopes with plants that will retain water and help reduce runoff. Group plants according to their watering needs.

- Apply a layer of mulch around trees and plants. Mulch slows evaporation of moisture while discouraging weed growth. Adding 5–10 centimeters (2–4 inches) of organic material such as compost or bark mulch will increase the soil's ability to retain moisture. Press down the mulch around the drip line of each plant to form a slight depression, which will prevent or minimize water runoff.

- Avoid overwatering plants and shrubs. This can actually diminish plant health and cause yellowing of the leaves.

- Many beautiful shrubs and plants can thrive with far less watering than other species. Replace herbaceous perennial borders with native plants, which use less water and are more resistant to local plant diseases. Consider applying the principles of xeriscaping for a low-maintenance, drought-resistant yard.

- Position your sprinklers so that the water lands on the lawn or garden, not on paved areas.

- Water your lawn only when it needs it. A good way to find out if your lawn needs watering is to step on the grass. If it springs back when you lift your foot, it doesn't need water. If it stays flat, the lawn is ready for watering. Letting the grass grow taller—to about 8–10 centimeters (3–4 inches)—will also promote water retention in the soil.

- Add organic matter to garden soil to help increase its absorption and water retention. Areas that are already planted can be top-dressed with compost or organic matter.

- Most lawns need only about 2.5 centimeters (1 inch) of water per week. During dry spells you can stop watering altogether: the lawn will turn brown and go dormant. Once cooler weather arrives, the morning dew and rainfall will bring the grass back to its usual vigor. This may result in a brown summer lawn but it saves a lot of water.

- Deep-soak your lawn. When watering, do it long enough for the moisture to soak down to the roots, where it will do the most good. A light sprinkling can evaporate quickly and tends to encourage shallow root systems. Put an empty tuna can on your lawn—when it's full, you've watered about the right length of time.

- Early or late watering reduces water loss due to evaporation. Early morning is generally better than dusk, since it helps prevent the growth of fungus, and watering early in the day is also the best defense against slugs and other garden pests. Try not to water when it's windy; wind can blow the water off target and speed evaporation.

- Use efficient watering systems for shrubs, flowerbeds and lawns. You can greatly reduce the amount of water used with soaker hoses or a simple drip irrigation system.

- Use a commercial drive-through car wash. Most car washes use recycled water for cleaning.

- If washing the car by hand, don't run the hose while washing. Clean the car first, using a pail of soapy water. Use the hose only for rinsing—this simple practice can save as much as 570 liters (150 gallons). Use a spray nozzle when rinsing for more efficient use of water.

- Sweep driveways and sidewalks clean instead of hosing them down.

- Check for leaks in outdoor pipes, hoses, faucets and couplings. Leaks outside the house may not seem so bad, since they're not as visible, but they can be just as wasteful as leaks indoors. Check frequently to keep them drip-free. Use washers at spigots and hose connections to eliminate leaks.

Lifestyle

- Drive less. In Alberta's oil sands, a barrel of oil takes about 2.5 barrels of water to produce, and most of the water used ends up in toxic tailings ponds because it's too polluted to return to the river. By driving less, you can become part of the solution to protect rivers such as the Athabasca.

- Create awareness of the need for water conservation among your children. Avoid purchasing recreational toys that require a constant stream of water.

- Be aware of and follow all water conservation and water shortage rules and restrictions that may be in effect in your area.

- Encourage your employer to promote water conservation in the workplace. Suggest that water conservation be made part of employee orientation and training.

- Encourage your school system and local government to help develop and promote a water conservation ethic among both children and adults.

- Support projects that will lead to increased use of reclaimed wastewater for irrigation and other uses.

- Report all significant water loss (broken pipes, open hydrants, errant sprinklers, abandoned free-flowing wells, etc.) to property owners, local authorities or public works departments.

- Support efforts and programs to create concern for water conservation among tourists and visitors to your area. Make sure visitors understand the need for and benefits of water conservation.

- Conserve water because it is the right thing to do. Don't waste water just because someone else is footing the bill, such as when you are showering at a health club or community center.

- Purchase secondhand clothing. The average water requirement to produce 1 kilogram (2 pounds) of cotton is 11,000 liters (2,900 gallons). This translates to a whopping 2,900 liters (766 gallons) to produce a plain cotton shirt! You can achieve big reductions in your water footprint by buying clothes secondhand or by wearing polyester, which requires much less water to produce.

- Patronize businesses that practice and promote water conservation.

- Encourage your friends and neighbors to be part of a water-conscious community. Promote water conservation in community newsletters, on bulletin boards and by example.

- Try to do one thing every day that will result in saving water. Don't worry if the savings are minimal. Every drop counts, and every person can make a difference. So tell your friends, neighbors and co-workers, "Turn it off and keep it off."

Sources

See page 131 for full-text references

page 7 - Hoekstra 2013; US Environmental Protection Agency 2013

page 8 - Hoekstra 2013; UNICEF/WHO, cited in World Business Council for Sustainable Development 2006

page 9 - United Nations World Water Assessment Programme 2012; US Geological Service; Hoekstra et al 2012; Global Water Intelligence (2014)

page 10 - US Geological Service; Hoekstra 2013; United States Environmental Protection Agency 2012; Brooymans 2011; United Nations 2010

page 11 - World Bank "Water-Energy Nexus"; International Energy Agency 2012; Hoekstra 2013; Lawrence Livermore National Laboratory/ Department of Energy cited by Lloyd Alter 2014; Ceres 2014; Kiger 2013; Payne 2014

page 13 - Struzik 2013; Gray 2013; Hoekstra 2013; National Public Radio 2012; Harvey and Kassai 2013; Lane 2013; UN World Water Assessment Programme 2012; International Energy Agency 2012 cited in Hoekstra 2013; Transports Québec 2013

page 14 - World Bank 2008; International Energy Agency 2012; World Economic Forum 2011; Union of Concerned Scientists 2010; United States Energy Information Agency; Phillips 2014

page 15 - American Wind Energy Association 2014; Sustainable Energy For All 2013; United Nations Environment Program 2014; Jacobson 2014; Hoekstra 2013

page 16 - DiBlasio 2014; Pacific Institute 2012; Hoekstra 2013; United Nations Development Programme 2006a

page 18 - Hoekstra 2013; Water Footprint Network; Salzman 2013; Container Recycling Institute 2013; Pacific Institute 2009; Natural Resources Defense Council revised 2013

page 19 - Hoekstra 2013

page 20 - Hoekstra 2013; Earth Policy Institute 2012; US Environmental Protection Agency; Natural Resources Defense Council 2012; Kostigan 2010

page 21 - Ravasio 2012; World Wildlife Fund; Organic Trade Association 2012; Dishman 2013; Hoekstra 2013

page 24 - Shiklomanov and Rodda 2003

page 25 - Shiklomanov and Rodda 2003, cited in Black and King 2009; Environment Canada 2009

page 26 - Balliett 2010

page 27 - Habermann, Moen and Stykel 2012

page 28 - Strassman 2010

page 29 - Black 2004; Brooymans 2011

page 30 - Shiklomanov and Rodda 2003 cited in Black and King 2009; Balliett 2010

page 31 - Kromm and White 1992; Workman 2010

page 32 - Foster, Lawrence and Morris 1998; World Business Council for Sustainable Development 2006; Black and King 2009; Haugen and Musser 2012; Qiu 2010

page 33 - Rooney 2009; Spilsbury 2010

page 34 - Black and King 2009; Barlow 2007

page 35 - UN FAO Aquastat Database (2011 figures)

pages 36–37 - cited in United Nations World Water Assessment Programme 2012

page 38 - Stockdale, Sauter and McIntyre 2010

page 39 - FAO, cited in World Business Council for Sustainable Development, 2006.

page 40 - Balliett 2010; Bell 2009

page 41 - McMurdoDryValleys.ag

page 42 - Black 2004

page 43 - Balliett 2010

page 46 - OHCHR, cited in UN–Water Decade Programme on Advocacy and Communication 2012

page 47 - Black 2004; UNICEF & WHO, cited in World Business Council for Sustainable Development 2006

page 48 - Kostigan 2010

page 49 - Proctor and Gamble Company, in Appropedia 2013; Green Mama 2013

page 50 - Fishman 2011; Gasson 2008; Environment Canada 2012

page 51 - Brooymans 2011; Haugen and Musser 2012; Union of Concerned Scientists 2010

page 52 - Natural Resources Canada 2006

page 53 - Black and King 2009; Worldwatch Institute 2004

page 54 - Environment Canada 2011

page 55 - U.S. Geological Service 2009

page 56 - Fishman 2011; International Bottled Water Association 2013

page 57 - Consumer Reports 2008; Waterwise 2011

page 58 - U.S. Environmental Protection Agency 2013; U.S. Geological Service 2013

page 59 - Mekonnen and Hoekstra 2010a; Chapagain, Hoekstra et al. 2005

page 60 - Fishman 2011; Levi Strauss & Co. 2009

page 61 - Cope 2009; Fishman 2011; Shiva 2002

page 62 - International Telecommunication Union 2013; Oki 2010

page 63 - Hoekstra and Chapagain 2007; Mekonnen and Hoekstra 2010b

page 64 - Hoesktra 2013; Mekonnen, Hoekstra and Becht 2012

page 65 - DeBeers 2010

page 68 - Health Canada 2012

page 69 - Environmental Working Group 2009

page 70 - Chapagain and Hoekstra 2003

page 71 - Chapagain and Hoekstra 2003

page 72 - Hoekstra and Mekonnen 2010; Kostigan 2010

page 73 - Hoekstra and Mekonnen 2010

page 74 - FAO 2011; Chapagain and Orr 2008

page 75 - FAO 2011; Chapagain and Orr 2008

page 76 - Hoekstra and Chapagain 2008; Mekonnen and Hoekstra 2010

page 77 - Hoekstra 2012; Hoekstra and Mekonnen 2010; Kostigan 2010

pages 78–79 - Allan 2011; A. Y. Hoekstra 2012

page 80 - Aldaya and Hoekstra 2010

page 81 - Chapagain and Hoekstra 2004

page 82 - Mekonnen and Hoekstra 2010b

page 83 - Mekonnen and Hoekstra 2010b; Kostigan 2010

page 84 - Aldaya and Hoekstra 2010; Ruini, et al. 2013

page 85 - Hoekstra 2013

page 86 - Kostigan 2010

page 87 - Hoekstra and Mekonnen 2010

page 88 - Mekonnen and Hoekstra 2010a, Kostigan 2010

page 89 - Mekonnen and Hoekstra 2010a; Thirlwall 2011

page 90 - Nature Conservancy/Coca-Cola Company 2010; Workman 2010

page 91 - Mekonnen and Hoekstra 2010a; Kostigan 2010

page 94 - National Round Table on the Environment and the Economy 2010

page 95 - Gerbens-Leenes et al. 2012; Mekonnen and Hoekstra 2010a; Dominguez-Faus et al. 2009

page 96 - Dominguez-Faus et al. 2009; United Nations World Water Assessment Programme 2012

page 97 - Kostigan 2010

page 98 - SAE International 2012

page 99 - Black and King 2009; Balliett 2010

pages 100–101 - Hoekstra 2013

page 102 - Balliett 2010

page 103 - Bell 2009; Siebert et al. 2010

page 104 - United Nations World Water Assessment Programme 2012

page 105 - Statistics Canada, cited in National Round Table on the Environment and the Economy 2010

page 106 - Black and King 2009; United Nations World Water Assessment Programme 2012

page 107 - Kostigan 2010

page 108 - Rogers and Leal 2010

page 109 - Balliett 2010

page 110 - Imperial County Farm Bureau 2008; University of California 2011

page 111 - U.S. Geological Service 2009

page 113 - Hoekstra 2013; World Economic Forum 2011; Brown 2012

page 114 - Government of Canada; US National Climate Assessment 2014; Owens 2014; Magill 2013; Carlson 2014; Romm 2014; California Department of Agriculture; Gillam 2013; Brown 2012; Pacific Institute 2012

page 115 - Rogers and Fleur 2014; Environmental Management 2005 cited in Diep 2011; Rogers 2014; Hoekstra 2013

page 116 - World Trade Organization 2013; Barlow 2013; Hoekstra 2013; National Geographic; Sidahmed 2013; Leahy 2013; Wissing 2014; United States Department of Agriculture 2013; Hoekstra 2013; Vasil 2013; de Châtel 2014; Pacific Institute 2012

page 117 - Barlow 2013

References

Aldaya, M.M., and A.Y. Hoekstra. (2010). "The Water Needed for Italians to Eat Pasta and Pizza." Agricultural Systems 103, no. 6 (July): 351–60. doi:10.1016/j.agsy.2010.03.004.

Allan, Tony. (2011). Virtual Water: Tackling the Threat to Our Planet's Most Precious Resource. London: I.B. Tauris.

Alter, Lloyd. (2014). "Latest look at the Lawrence Livermore graph that tells you everything you need to know about America's energy use" http://www.treehugger.com/energy-efficiency/latest-look-lawrence-livermore-graph-tells-you-everything-you-need-know-about-americas-energy-use.html

American Wind Energy Association. (2014). "Wind energy secures significant CO2 emission reductions for the U.S" http://www.awea.org/MediaCenter/pressrelease.aspx?ItemNumber=6320

Âpihtawikosisân. (2012). "Dirty Water, Dirty Secret." Apihtawikosisan.com, November 8. http://apihtawikosisan.com/2012/11/08/dirty-water-dirty-secret-full-article (accessed October 16, 2013).

Appropedia. (2013). "Life Cycle Assessment: Cloth vs. Disposable Diapers." Appropedia.org, April 22. http://www.appropedia.org/Cloth_versus_disposable_diapers (accessed October 16, 2013).

Balliett, James Fargo. (2010). Freshwater: Environmental Issues, Global Perspectives. Armonk, NY: Sharpe Reference.

Barlow, Maude. (2007). Blue Covenant: The Global Water Crisis and the Coming Battle for the Right to Water. Toronto: McClelland & Stewart.

Barlow, Maude. (2013). Blue Future: Protecting Water for People and the Planet Forever. Toronto: House of Anansi Press

Barlow, Maude, and Tony Clarke. (2002). Blue Gold: The Fight to Stop the Corporate Theft of the World's Water. New York: W.W. Norton.

Bell, Alexander. (2009). Peak Water: Civilization and the World's Water Crisis. Edinburgh: Luath Press.

Berga, Luis. (2006). Dams and Reservoirs, Societies and Environment in the 21st Century. London: Taylor & Francis.

Black, Maggie. (2004). The No-Nonsense Guide to Water. Oxford: New Internationalist.

Black, Maggie, and Jannet King. (2009). The Atlas of Water. 2nd ed. Berkeley: University of California Press.

Brazzale SpA. (2012). "Gran Moravia Is the First Cheese in the World to Set Its Water Footprint." Brazzale.com, September 26. http://www.brazzale.com/gran-moravia-is-the-first-cheese-in-the-world-to-set-its-water-footprint (accessed October 7, 2013).

Brooymans, Hanneke. (2011). Water in Canada: A Resource in Crisis. Edmonton: Lone Pine.

Brown, Lester. (2012). Full Planet, Empty Plates: The New Geopolitics of Food Scarcity. W. W. Norton & Company.

California Department of Agriculture. "CALIFORNIA AGRICULTURAL PRODUCTION STATISTICS" http://www.cdfa.ca.gov/statistics/ Campbell, Mary K., and Shawn O. Farrell. (2007). Biochemistry. 6th ed. Belmont, CA: Cengage Learning.

Canadian Association of Petroleum Producers. (2012). "Water Use in Canada's Oil Sands." Calgary, AB: CAPP.

Carlson, Kathryn Blaze. (2014). "Water level drop expected to hit property values on Great Lakes" Global and Mail. http://www.theglobeandmail.com/report-on-business/industry-news/energy-and-resources/great-lakes-facing-a-water-level-crisis/article19344789/

Ceres. (2014.) "Hydraulic Fracturing & Water Stress" https://www.ceres.org/resources/reports/hydraulic-fracturing-water-stress-water-demand-by-the-numbers/view

Chapagain, A.K., and A.Y. Hoekstra. (2003). The Water Needed to Have the Dutch Drink Tea. Value of Water Research Reports, series 15. Delft: UNESCO-IHE.

—. (2004). Water Footprints of Nations. Value of Water Research Reports, series 16. Delft: UNESCO-IHE.

—. (2007). "The Water Footprint of Coffee and Tea Consumption in the Netherlands." Ecological Economics 64, no. 1: 109–18.

—. (2010). The Green, Blue and Grey Water Footprint of Rice from Both a Production and Consumption Perspective. Value of Water Research Reports, series 40. Delft: UNESCO-IHE.

Chapagain, A.K., A.Y. Hoekstra, H.H.G. Savenije and R. Gautam. (2005). The Water Footprint of Cotton Consumption. Value of Water Research Reports, series 18. Delft: UNESCO-IHE.

Chapagain, A.K., and S. Orr. (2008). UK Water Footprint: The Impact of the UK's Food and Fibre Consumption on Global Water Resources. Vol. 1. Surrey, UK: World Wildlife Federation.

Christensen, Randy. (2001). Waterproof: Canada's Drinking Water Report Card. Toronto: Sierra Legal Defence Fund.

—. (2011). Waterproof 3: Canada's Drinking Water Report Card. Vancouver: Ecojustice.

Consumer Reports. (2008). "Shower or Bath: Which Uses More Water?" ConsumerReports.org, August 22. http://news.consumerreports.org/home/2008/08/shower-or-bath.html (accessed July 7, 2013).

Container Recycling Institute. (2013). "Bottled Up (2000-2010) - Beverage Container Recycling Stagnates" http://www.container-recycling.org/index.php/publications/2013-bottled-up-report

Cope, Gord. (2009). "Pure Water, Semiconductors and the Recession." Global Water Intelligence 10, no. 10 (October).

DeBeers. (2010). "Living Up to Diamonds: Report to Society 2010." http://www.debeersgroup.com/ImageVaultFiles/id_989/cf_5/RTS_10_Full.pdf (accessed October 31, 2013).

De Châtel, Francesca. (2014). " The Role of Drought and Climate Change in the Syrian Uprising: Untangling the Triggers of the Revolution". Middle Eastern Studies, DOI: 10.1080/00263206.2013.850076

De Villiers, Marcus. (1999). Water: The Fate of Our Most Precious Resource. Toronto: Stoddart.

Diani, Hera. (2009). "The Sewage: Poor Sanitation Means Illness and High Costs." Jakarta Globe, July 24. http://www.thejakartaglobe.com/waterworries/the-sewage-poor-sanitation-means-illness-and-high-costs/320019 (accessed October 31, 2013).

DiBlasio, Natalie. (2014). "Cities spend billions to keep sewage out of your rivers" USA TODAY http://www.usatoday.com/story/news/nation/2014/03/09/sewer-overflow-tunnel/5808615/

Dishman, Lydia. (2013). "Inside H&M's Quest For Sustainability In Fast Fashion" Forbes http://www.forbes.com/sites/lydiadishman/2013/04/09/inside-hms-quest-for-sustainability-in-fast-fashion/

Diep, Francie. (2011). "Lawns vs. crops in the continental U.S." Science Line http://scienceline.org/2011/07/lawns-vs-crops-in-the-continental-u-s/ Dominguez-Faus, R., Susan E. Powers, Joel G. Burken and Pedro J. Alvarez. (2009). "The Water Footprint of Biofuels: A Drink or Drive Issue?" Environmental Science & Technology 43, no. 9 (May): 3005–10. doi:10.1021/es802162x.

Earth Policy Institute. (2012). "Peak Meat: U.S. Meat Consumption Falling" http://www.earth-policy.org/data_highlights/2012/highlights25

Environmental Working Group. (2009). "Over 300 Pollutants in US Tap Water." National Drinking Water Database. December. http://www.ewg.org/tap-water/ (accessed October 31, 2013).

Environment Canada. (2009a). "How Much Do We Have?" Resources, November 19. http://www.ec.gc.ca/eau-water/default.asp?lang=En&n=51E3DE0C-1 (accessed October 31, 2013).

—. (2009b). "Water: No Time to Waste – A Consumer's Guide to Water Conservation." Resources, November 26. http://www.ec.gc.ca/eau-water/default.asp?lang=en&n=344B115B-1 (accessed October 31, 2013).

—. (2011). "Groundwater." Resources, February 15. http://www.ec.gc.ca/eau-water/default.asp?lang=En&n=300688DC-1 (accessed October 31, 2013).

—. (2012). "Withdrawal Uses." Resources, June 13. http://www.ec.gc.ca/eau-water/default.asp?lang=En&n=851B096C-1 (accessed April 15, 2013).

FAO Water Development and Management Unit. (2011). "Crop Water Information: Tomato." Natural Resources and Environment Department. November 25. http://www.fao.org/nr/water/cropinfo_tomato.html (accessed October 31, 2013).

Fishman, Charles. (2011). The Big Thirst: The Secret Life and Turbulent Future of Water. New York: Free Press.

Foster, Stephen, Adrian Lawrence and Brian Morris. (1998). Groundwater In Urban Development: Assessing Management Needs and Formulating Policy Strategies. Washington, DC: World Bank.

Gasson, Christopher. (2008). "World Water Prices Rise by 6.7%." Global Water Intelligence 9, no. 9 (September).

Gerbens-Leenes, P.W., A.R. van Lienden, A.Y. Hoekstra and Th.H. van der Meer. (2012). "Biofuel Scenarios in a Water Perspective: The Global Blue and Green Water Footprint." Global Environmental Change 22: 764–75.

Gillam, Carey. (2013). "Ogallala aquifer: Could critical water source run dry?" Reuters. http://www.csmonitor.com/Environment/Latest-News-Wires/2013/0827/Ogallala-aquifer-Could-critical-water-source-run-dry Gleick, Peter. (2011). The World's Water. Vol. 7: The Biennial Report on Freshwater Resources. Washington, DC: Island Press.

Global Water Intelligence (2014). "Thirsty energy: the conflict between demands for power and water" http://www.theguardian.com/sustainable-business/thirsty-energy-conflict-energy-demand-water-access

Government of Canada. "Canada's Action on Climate Change" http://www.climatechange.gc.ca/default.asp?lang=en&n=036D9756-1 The Green Mama. (2013). "Analyzing Environmental Life-Cycle Costs of Diapers." http://www.thegreenmama.com/analyzing-environmental-life-cycle-costs-diapers (accessed October 31, 2013).

Gray, Tim. (2014) "The tar sands don't have to pollute the water. So why do they?" Global and Mail http://www.theglobeandmail.com/globe-debate/the-tar-sands-dont-have-to-pollute-the-water-so-why-do-they/article14564780/

Habermann, R., S. Moen and E. Stykel. (2012). Superior Facts. Duluth, MN: Minnesota Sea Grant.

Hammond, Michael. (2013). The Grand Ethiopian Renaissance Dam and the Blue Nile: Implications for Transboundary Water Governance. GWF discussion paper 1307. Canberra: Global Water Forum.

Harvey, Christine and Lucia Kassai. (2013). "Ethanol Falls as EPA Considers Scaling Back U.S. Mandate" Bloomberg. http://www.bloomberg.com/news/2013-10-11/ethanol-falls-as-epa-considers-scaling-back-u-s-mandate.html

Haugen, David, and Susan Musser. (2012). Will the World Run Out of Fresh Water? Detroit: Greenhaven Press.

Health Canada. (2012). "Canadian Drinking Water Guidelines." Environmental and Workplace Health. November 11. http://www.hc-sc.gc.ca/ewh-semt/water-eau/drink-potab/guide/index-eng.php (accessed October 31, 2013).

—. (2013). "Drinking Water and Wastewater." First Nations and Inuit Health. February 26. http://www.hc-sc.gc.ca/fniah-spnia/promotion/public-publique/water-eau-eng.php (accessed October 31, 2013).

Hoekstra, A.Y. (2012). "The Hidden Water Resource Use Between Meat and Dairy." Animal Frontiers 2, no. 2

Hoekstra, A.Y. (2013): "The Water Footprint of Modern Consumer Society", Routledge.

—. (2013). The Water Footprint of Modern Consumer Society. New York: Routledge.

Hoekstra, A.Y, and A.K. Chapagain. (2007). "Water Footprints of Nations: Water Use by People as a Function of Their Consumption Pattern," Water Resources Management 21, no. 1 (January): 35–48. doi:10.1007/s11269-006-9039-x.

—. (2008). Globalization of Water: Sharing the Planet's Freshwater Resources. Oxford: Blackwell.

Hoekstra, A.Y., A.K. Chapagain, M.M. Aldaya and M.M. Mekonnen. (2011). The Water Footprint Assessment Manual: Setting the Global Standard. London: Earthscan.

Hoekstra, A.Y., and M.M. Mekonnen. (2011). National Water Footprint Accounts: The Green, Blue and Grey Water Footprint of Production and Consumption. Value of Water Research Reports, series 50. Delft: UNESCO-IHE.

Hoekstra, A.Y., M.M. Mekonnen, A.K. Chapagain, R.E. Matthews, B.D. Richter. (2012) "Global Monthly Water Scarcity: Blue Water Footprints versus Blue Water Availability" PLOS one, February 29. DOI: 10.1371/journal.pone.0032688.

Hongqiao, Liu. (2012). "Stormy Weather on Cloud-Seeding." Caixin Online, August 13. http://english.caixin.com/2012-08-13/100423557.html (accessed October 31, 2013).

Ibrahim, Abadir M. (2011). "The Nile Basin Cooperative Framework Agreement: The Beginning of the End of Egyptian Hydro-political Hegemony." Missouri Environmental Law and Policy Review 18.

Imperial County Farm Bureau. (2008). "Quick Facts about Imperial County Agriculture." http://www.icfb.net/countyag.html (accessed November 22, 2013).

International Bottled Water Association. (2013). "2011 Market Report Findings." Statistics, April. http://www.bottledwater.org/files/2011BWstats.pdf (accessed October 31, 2013).

International Energy Agency. (2012). World Energy Outlook 2012. http://www.worldenergyoutlook.org/resources/water-energynexus/

International Standards Organization. (2013). "No More Waste: Tracking Water Footprints." July 29. http://www.iso.org/iso/home/news_index/news_archive/news.htm?refid=Ref1760 (accessed December 4, 2013).

International Telecommunication Union. (2013). "Mobile Subscriptions Near the 7 Billion Mark: Does Almost Everyone Have a Phone?" ITU News 2.

Jacobson, Mark. (2014). "Plans to Convert the 50 United States to Wind, Water, and Sunlight" http://web.stanford.edu/group/efmh/jacobson/Articles/I/WWS-50-USState-plans.html

Kiger, Patrick J. (2013). "North Dakota's Salty Fracked Wells Drink More Water to Keep Oil Flowing". National Geographic News. http://news.nationalgeographic.com/news/energy/2013/11/131111-north-dakota-wells-maintenance-water/

Kostigen, Thomas M. (2010). The Green Blue Book. Emmaus, PA: Rodale.

Kromm, David E., and Stephen E. White. (1992). Groundwater Exploitation in the High Plains. Lawrence: University Press of Kansas.

Lane, Jim. (2013). "Biofuels Mandates Around the World: 2014" Biofuels Digest. http://www.biofuelsdigest.com/bdigest/2013/12/31/biofuels-mandates-around-the-world-2014/

Leahy, Stephen. (2013). "U.S., Malaysia Lead Worldwide "Land Grabs"". Inter Press News. http://www.ipsnews.net/2013/06/u-s-malaysia-lead-worldwide-land-grabs/

Levi Strauss & Co. (2009). "A Product Life Cycle Approach to Sustainability." March. http://www.levistrauss.com/library/product-lifecycle-approach-sustainability (accessed October 31, 2013).

Magill, Bobby. (2103). "Is the West's Dry Spell Really a Megadrought?" Climate Central. http://www.climatecentral.org/news/is-the-wests-dry-spell-really-a-megadrought-16824

Mekonnen, M.M., and A.Y. Hoekstra. (2010a). "The Green, Blue and Grey Water Footprint of Crops and Derived Crop Products." Hydrology and Earth System Sciences 15, no. 5: 1577–1600. doi:10.5194/hess-15-1577-2011.

—. (2010b). The Green, Blue and Grey Water Footprint of Farm Animals and Animal Products. Value of Water Research Reports, series 48. Delft: UNESCO-IHE.

Mekonnen, M.M., A.Y. Hoekstra and R. Becht. (2012). "Mitigating the Water Footprint of Export Cut Flowers from the Lake Naivasha Basin, Kenya." Water Resource Management 26: 3725–42. doi:10.1007/s11269-012-0099-9.

National Geographic. "Saudi Arabia's Great Thirst" http://environment.nationalgeographic.com/environment/freshwater/saudi-arabia-water-use/

National Public Radio. (2012). "Congress Ends Era Of Ethanol Subsidies" http://www.npr.org/2012/01/03/144605485/congress-ends-era-of-ethanol-subsidiesm%5D

National Round Table on the Environment and the Economy. (2010). Changing Currents: Water Sustainability and the Future of Canada's Natural Resource Sectors. Ottawa: Government of Canada (NRTEE).

Natural Resources Canada. (2006). "2003 Survey of Household Energy Use (SHEU): Detailed Statistical Report." Office of Energy Efficiency. http://oee.nrcan.gc.ca/publications/statistics/sheu03/pdf/sheu03.pdf (accessed October 31, 2013).

Natural Resources Defense Council. (Revised 2013). "Bottled Water

— "Atlas of Water". http://atlas.gc.ca/site/english/maps/water.html

"Pure Drink or Pure Hype?" http://www.nrdc.org/water/drinking/bw/chap1.asp#note33

— (2012). "Wasted: How America Is Losing Up to 40 Percent of Its Food from Farm to Fork to Landfill" http://www.nrdc.org/food/files/wasted-food-ip.pdf

The Nature Conservancy and Coca-Cola Company. (2010). "Product Water Footprint Assessments: Practical Application in Corporate Water Stewardship." September. http://www.thecoca-colacompany.com/presscenter/

TCCC_TNC_WaterFootprintAssessments.pdf (accessed October 31, 2013).

Netherwood, Marshall. (2008). "Water Allocations and Use: Alberta Oil and Gas Industry." University of Saskatchewan IP3 Center for Hydrology. March. http://www.usask.ca/ip3/download/canmore2008/presentations/

Netherwood.pdf (accessed October 31, 2013).

Oki, Taikan. (2010). "Issues on Water Footprint and Beyond." Institute of Industrial Science, University of Tokyo. June 3. http://gec.jp/gec/en/Activities/ietc/fy2010/wf/wf_os-3e.pdf (accessed October 31, 2013).

Organic Trade Association. (2012). "Organic Cotton Facts" http://www.ota.com/organic/fiber/organic-cotton-facts.html

Owens, Caitlin. (2014). "California's drought getting even worse, experts say" Los Angles Times http://www.latimes.com/local/lanow/la-me-ln-drought-worsens-across-california-20140619-story.html

Pacific Institute. (2007). "Bottled Water and Energy: A Fact Sheet." February. http://www.pacinst.org/topics/

— 2009. "Energy Implications of Bottled Water" http://pacinst.org/publication/energy-implications-of-bottled-water/

— 2012. "Assessment of California's Water Footprint" http://www2.pacinst.org/publication/assessment-of-californias-water-footprint/

— 2012. Water Conflict Chronology List. http://www2.worldwater.org/conflict/list/water_and_sustainability/bottled_water/bottled_water_and_energy.html (accessed October 31, 2013).

Payne, Heather. NEXUS 2014: WATER, FOOD, CLIMATE AND ENERGY CONFERENCE, MARCH 5-8, 2014 University of North Carolina.

Phillips, Ari. (2014). "The Ivanpah Solar Power Plant Uses Relatively Little Water". Clean Technica http://cleantechnica.com/2014/02/15/ivanpah-solar-power-plant-uses-relatively-little-water/

Polaris Institute. (2009). "Murky Waters: The Urgent Need for Health and Environmental Regulations of the Bottled Water Industry." Ottawa: Polaris Institute.

Pushak, Nataliya, and Vivien Foster. (2011). Angola's Infrastructure: A Continental Perspective. Policy Research Working Paper 5813. Washington, DC: World Bank.

Qiu, Jane. (2010). "China Faces Up to Groundwater Crisis." Nature 466, no. 7304 (2010): 308. doi:10.1038/466308a.

Ravasio, Pamela. (2012). "How can we stop water from becoming a fashion victim?" The Guardian http://www.theguardian.com/sustainable-business/water-scarcity-fashion-industry

Rogers, Paul and Nicholas St Fleur. (2014). "California Drought: Database shows big difference between water guzzlers and sippers" San Jose Mercury News. http://www.mercurynews.com/science/ci_25090363/california-drought-water-use-varies-widely-around-state%5D

Rogers, Paul. (2014). "California drought: Tips for conserving water" San Jose Mercury News. http://www.mercurynews.com/science/ci_24952670/california-drought-tips-conserving-water

Rogers, Peter, and Susan Leal. (2010). Running Out of Water: The Looming Crisis and Solutions to Conserve Our Most Precious Resource. New York: Palgrave MacMillan.

Romm, Joe. (2014). "Leading Scientists Explain How Climate Change Is Worsening California's Epic Drought" Climate Progress. http://thinkprogress.org/climate/2014/01/31/3223791/climate-change-california-drought/

Rooney, Anne. (2009). Sustainable Water Resources. Minneapolis: Arcturus.

Royal Bank of Canada. (2013). "RBC Canadian Water Attitudes Study." About RBC, March 20. http://www.rbc.com/community-sustainability/environment/rbc-blue-water/water-attitude-study.html (accessed October 31, 2013).

Ruini, L., M. Marino, S. Pignatelli, F. Laio and L. Ridolfi. (2013). "Water Footprint of a Large-Sized Food Company: The Case of Barilla Pasta Production." Water Resources and Industry 1/2 (March–June): 7–24.

SAE International. (2012). Quantifying the Life Cycle Water Consumption of a Passenger Vehicle. January. 2012-01-646. SAE International.

Salzman, James. (2013). Drinking Water. Overlook Duckworth

Shiklomanov, I.A., and John C. Rodda. (2003). World Water Resources at the Beginning of the 21st Century. Cambridge: UNESCO/Cambridge University Press.

Shiva, Vandana. (2002). Water Wars: Privatization, Pollution and Profit. Toronto: Between the Lines.

Sidahmed, Alsir. (2013). "Saudi moves mirror Arab food security issues". Arab News. http://www.arabnews.com/news/450468

Siebert, S., et al. (2010). "Groundwater Use for Irrigation: A Global Inventory." Hydrology and Earth System Sciences 14: 1863–80. doi:10.5194/hess-14-1863-2010.

Soil & More International BV. (2011). "Water Footprint Assessment: Bananas and Pineapples." Dole Corporate Responsibility and Sustainability. http://dolecrs.com/uploads/2012/06/Soil%20&%20More%20Water%20Footprint%20Assessment.pdf (accessed November 1, 2013).

Spilsbury, Louise. (2010). Threats to Our Water Supply: Can The Earth Survive? New York: Rosen.

Stockdale, Charles B., Michael B. Sauter and Douglas A. McIntyre. (2010). "The Ten Biggest American Cities That Are Running Out of Water." 24/7 Wall Street, October 29. http://247wallst.com/investing/2010/10/29/the-ten-great-american-cities-that-are-dying-of-thirst/ (accessed December 3, 2013).

Strassman, Mark. (2010). "America's Dwindling Water Supply." CBS Evening News, January 9. http://www.cbsnews.com/stories/2010/01/08/eveningnews/main6073416.shtml (accessed November 1, 2013).

Struzik, Ed. (2013) "With Tar Sands Development, Growing Concern on Water Use." Yale 360 http://e360.yale.edu/feature/with_tar_sands_development_growing_concern_on_water_use/2672/

Sustainable Energy For All. (2013) "Universal Energy Access" http://www.se4all.org/our-vision/our-objectives/universal-energy/

Swain, Ashok. (2011). "Challenges for Water Sharing in the Nile Basin: Changing Geo-politics and Changing Climate." Hydrological Sciences Journal 56, no. 4: 687–702. doi:10.1080/02626667.2011.577037.

Thirlwall, Claire. (2011). "Is Water Use in the Chocolate Industry Excessive?" Urban Times, November 29. http://urbantimes.co/2011/11/is-water-use-in-the-chocolate-industry-excessive/ (accessed October 31, 2013).

Transports Québec. (2013). "Water Levels in the St. Lawrence River." Climate Change. http://www.mtq.gouv.qc.ca/portal/page/portal/ministere_en/ministere/environnement/changements_climatiques/adapter_transports_impacts_changements_climatiques/niveaux_eau_saint-laurent (accessed October 31, 2013).

United States Department of Agriculture. (2013). "Egypt." Foreign Agriculture Service. http://gain.fas.usda.gov/Recent%20GAIN%20Publications/Citrus%20Annual_Cairo_Egypt_12-12-2013.pdf

United States Energy Information Agency. (2012). FAQs http://www.eia.gov/tools/faqs/faq.cfm?id=97&t=3

United States Environmental Protection Agency. (2012). "The Great Lakes: An Environmental Atlas and Resource Book." June 25. http://www.epa.gov/greatlakes/atlas/index.html (accessed November 1, 2013).

—. (2013). "Indoor Water Use in the United States." WaterSense, June 20. http://www.epa.gov/watersense/pubs/indoor.html (accessed December 1, 2013).

United States Fish and Wildlife Service. (2013). "Species Reports." Environmental Conservation Online System. http://ecos.fws.gov/tess_public/SpeciesReport.do?groups=F&listingType=L&mapstatus=1 (accessed November 1, 2013).

United States Geological Service. (2009). Estimated Use of Water in the United States in 2005. USGS circular 3144. Reston, VA: US Geological Service.

—. (2013). "Water Properties and Measurements." USGS Water Science School. March 14. http://ga.water.usgs.gov/edu/waterproperties.html (accessed November 1, 2013).

U.S. Global Change Research Program. (2014). Third National Climate Assessment http://nca2014.globalchange.gov

Union of Concerned Scientists. (2010). "How It Works: Water for Power Plant Cooling." Clean Energy. http://www.ucsusa.org/clean_energy/our-energy-choices/energy-and-water-use/water-energy-electricity-cooling-power-plant.html (accessed November 1, 2013).

United Nations. (2010a). "General Assembly Adopts Resolution Recognizing Access to Clean Water, Sanitation." GA/10967, July 28. http://www.un.org/News/Press/docs/2010/ga10967.doc.htm (accessed November 1, 2013).

—. (2010b). "Resolution Adopted by the General Assembly on 28 July 2010. 64/292: The Human Right to Water and Sanitation." August 3. http://www.un.org/ga/search/view_doc.asp?symbol=A/RES/64/292 (accessed November 1, 2013).

United Nations Development Programme. (2006a). Human Development Report 2006. Beyond Scarcity: Power, Poverty and the Global Water Crisis. Geneva: UNDP.

—. (2006b). Niger Delta Human Development Report. Geneva: UNDP.

United Nations Water Decade Programme on Advocacy and Communication. (2012). The Human Right to Water and Sanitation. Geneva: UNW-DPAC/WSSCC.

United Nations World Water Assessment Programme. (2012). World Water Development Report 4: Managing Water under Uncertainty and Risk. Geneva: UNESCO-WWAP.

University of California. (2011). "Imperial County Agriculture." University of California Cooperative Extension. ceimperial.ucanr.edu/files/96429.pdf (accessed November 22, 2013).

UPM-Kymmene Corporation. (2011). From Forest to Paper: The Story of Our Water Footprint. Helsinki: UPM-Kymmene.

US Environmental Protection Agency. "Water Sense" http://www.epa.gov/WaterSense/pubs/indoor.html

US Geological Survey. "How much water is there on, in, and above the Earth" http://ga.water.usgs.gov/edu/earthhowmuch.html

van Oel, P.R., and A.Y. Hoekstra. (2012). "Towards Quantification of the Water Footprint of Paper: A First Estimate of Its Consumptive Component." Water Resource Management 26: 733–49. doi:10.1007/s11269-011-9942-7.

Vasil, Andrea. (2013). "Who's Water? Our Water" NOW. http://www.nowtoronto.com/columns/ecoholic.cfm?content=193051

Water Footprint Network. "Product Gallery" http://www.waterfootprint.org/?page=files/productgallery

Waterwise. (2011). "Showers vs. Baths: Facts, Figures and Misconceptions." News, November 24. http://www.waterwise.org.uk/news.php/11/showers-vs.-baths-facts-figures-and-misconceptions (accessed November 1, 2013).

Wissing, Carolyn. (2014). "Land grabbing: the real cost of buying cheap". Deutche Welle. http://www.dw.de/land-grabbing-the-real-cost-of-buying-cheap/a-17151695

Wolf, Aaron T., Shira B. Yoffe and Mark Girodano. (2003). "International Waters: Identifying Basins at Risk." Water Policy 5: 29–60.

Workman, James G. (2010). Diminishing Resources: Water. Greensboro, NC: Morgan Reynolds.

World Bank Databank. http://databank.worldbank.org/data/home.aspx

— 2013 "Thirsty Energy: Securing Energy in a Water-Constrained World". http://www.worldbank.org/en/topic/sustainabledevelopment/brief/water-energy-nexus

—2008 "A note on rising food prices" http://econ.worldbank.org/external/default/main?pagePK=64165259&piPK=64165421&theSitePK=469372&menuPK=64166093&entityID=000020439_20080728103002

World Business Council for Sustainable Development. (2006). Water Facts and Trends. Geneva: World Business Council for Sustainable Development.

World Economic Forum. (2011). Water Security. Island Press

Worldwatch Institute. (2004). "Matters of Scale: Planet Golf." Worldwatch Magazine 17, no. 2 (March/April). http://www.worldwatch.org/node/797 (accessed October 31, 2013).

— (2010) "Study Finds Rich U.S. Energy-Efficiency Potential" http://www.worldwatch.org/node/6212

World Trade Organization. (2013). "International Trade Statistics 2013" http://www.wto.org/english/res_e/statis_e/its2013_e/its2013_e.pdf

World Wildlife Fund. "Cotton: a water wasting crop" http://wwf.panda.org/about_our_earth/about_freshwater/freshwater_problems/thirsty_crops/cotton/

Photography Credits

Front cover background © Andrey Myagkov / Shutterstock; front cover top left © alexsl / iStock; front cover bottom, left to right © Nik Merkulov / Shutterstock, Bozena Fulawka / Shutterstock, Mazzzur / Shutterstock, Lyudmila Suvorova / Shutterstock; back cover left © AGCuesta / Shutterstock; back cover top © Gordana Sermek / Shutterstock; back cover middle © Indigolotos / Shutterstock; back cover bottom © Africa Studio / Shutterstock; page 2–3 © Kerenby / Shutterstock; page 6 © Leonello Calvetti/ Shutterstock; page 7 © Zffoto / Shutterstock; page 8 top © Meunierd / Shutterstock; page 8 middle © Chris Bence / Shutterstock; page 8 bottom © Alexander Kirch / Shutterstock; page 9 © / Shutterstock; page 10 © Calvindexter / Shutterstock; page 12 © Anekoho / Shutterstock; page 13 © Anekoho / Shutterstock; page 16 © Nolte Lourens / Shutterstock; page 17 © / Shutterstock; page 19 © Tyler Olson / Shutterstock; page 22 © Siro46 / Shutterstock; page 24 © Harvepino / Shutterstock; page 25 top © Tarasyuk Igor / Shutterstock; page 25 bottom © Alexi Bradich / Shutterstock; page 28 © Jiawangkun / Shutterstock; page 29 top © Gopixgo / Shutterstock; page 29 bottom © SergeBertasiusPhotography / Shutterstock; page 30 © Andre Nantel / Shutterstock; page 32 top © Tutti Frutti / Shutterstock; page 32 bottom © Nerthuz / Shutterstock; page 33 © 2bears / Shutterstock; page 34 inset © Arfabita / Shutterstock; page 36 left © Calvindexter / Shutterstock; page 36 right © Andresr / Shutterstock; page 37 left © Ixpert / Shutterstock; page 37 right © / Shutterstock; page 37 bottom © Brocreative / Shutterstock; page 40 © Bildagentur Zoonar GmbH / Shutterstock; page 41 © Sergey Tarasenko / Shutterstock; page 42 © Daniel J. Rao / Shutterstock; page 43 © Bikeriderlondon / Shutterstock; page 44 © Siro46 / Shutterstock; page 46 top © Sarawut Aiemsinsuk / Shutterstock; page 46 middle © Tim Booth / Shutterstock; page 46 bottom © Rashevskyi Viacheslav / Shutterstock; page 48 top © Wuttichok Painichiwarapun / Shutterstock; page 48 bottom © s74 / Shutterstock; page 49 top © Przemyslaw Ceynowa / Shutterstock; page 49 bottom © Myotis / Shutterstock; page 50 © Tarasov / Shutterstock; page 51 top © Melinda Fawver / Shutterstock; page 51 left © Erashov / Shutterstock; page 51 middle left © Cobalt88 / Shutterstock; page 51 middle right © Nordling / Shutterstock; page 51 right © Elena Elisseeva / Shutterstock; page 52 top © Spectral-Design / Shutterstock; page 52 bottom © M.M. / Shutterstock; page 53 top © Cappi Thompson / Shutterstock; page 53 bottom © Rashevskyi Viacheslav / Shutterstock; page 56 © Africa Studio / Shutterstock; page 57 top © Piotr Marcinski / Shutterstock; page 57 bottom © Eduard Stelmakh / Shutterstock; page 58 top © Canonzoom / Shutterstock; page 58 bottom left © Tommaso79 / Shutterstock; page 58 bottom right © Rihardzz / Shutterstock; page 59 top © Borislav Bajkic / Shutterstock; page 59 bottom © Borislav Bajkic / Shutterstock; page 60 left © Mexrix / Shutterstock; page 60 middle © Africa Studio / Shutterstock; page 60 right © Gordana Sermek / Shutterstock; page 61 © Ekipaj / Shutterstock; page 62 © Nik Merkulov / Shutterstock; page 63 top © RG-vc / Shutterstock; page 63 middle © Severija / Shutterstock; page 63 right © DnD-Production.com / Shutterstock; page 64 background © Africa Studio / Shutterstock; page 64 inset © Quang Ho / Shutterstock; page 65 © Kae Deezign / Shutterstock; page 66 © Siro46 / Shutterstock; page 68 left © Laboko / Shutterstock; page 68 right © Andrey Kuzmin / Shutterstock; page 69 © Andrey Kuzmin / Shutterstock; page 70 left © Epitavi / Shutterstock; page 70 right © Aleksandra Pikalova / Shutterstock; page 71 © Africa Studio / Shutterstock; page 72–73 © Viktar Malyshchyts / Shutterstock; page 74 top © Bronwyn Photo / Shutterstock; page 74 bottom © G215 / Shutterstock; page 75 © Peter Zijlstra / Shutterstock; page 76 top © Jerry Lin / Shutterstock; page 76 left © Sergio Schnitzler / Shutterstock; page 76 middle left © Arbalet / Shutterstock; page 76 middle right © / Shutterstock; page 76 right © Seregam / Shutterstock; page 77 top © Dulce Rubia / Shutterstock; page 77 middle © StockPhotosArt / Shutterstock; page 77 bottom left © Chiyacat / Shutterstock; page 77 bottom middle left © Margouillat photo / Shutterstock; page 77 bottom middle right © Dulce Rubia / Shutterstock; page 77 bottom right © Dulce Rubia / Shutterstock; page 78 © Ronald Sumners / Shutterstock; page 79 © Ronald Sumners / Shutterstock; page 80 top © Oxana Denezhkina / Shutterstock; page 80 bottom left © Dan Kosmayer / Shutterstock; page 80 bottom middle © Kiboka / Shutterstock; page 80 bottom right © Louella938 / Shutterstock; page 81 top © Lyudmila Suvorova / Shutterstock; page 81 bottom left © Ronald Sumners / Shutterstock; page 81 bottom middle © Kesu / Shutterstock; page 81 bottom right © Vadim Petrov / Shutterstock; page 82 top © Tankist276 / Shutterstock; page 82 bottom © Yevgeniy11 / Shutterstock; page 83 © Robyn Mackenzie / Shutterstock; page 84 left top © Timmary / Shutterstock; page 84 left bottom © Mffoto / Shutterstock; page 84 right © Thomas Bethge / Shutterstock; page 85 left, top to bottom © Shutterstock and: Coprid; ZCW; Andrey Starostin; Elena Schweitzer; Kiboka; Ifong; page 85 right © Freer / Shutterstock; page 86 left and right © Aaron Amat / Shutterstock; page 87 top © Studio red city/ Shutterstock; page 87 bottom © Kaarsten / Shutterstock; page 88 top © Luis Carlos Jimenez del rio / Shutterstock; page 88 bottom © Jiang Hongyan / Shutterstock; page 89 © Prapass / Shutterstock; page 90 © Maradonna 8888 / Shutterstock; page 91, left to right © Shutterstock and: GorillaAttack; Nitr; Africa Studio; Evgeny Karandaev; Bergamont; page 92 © Siro46 / Shutterstock; page 94 © Matee Nuserm / Shutterstock; page 95 © HomeStudio / Shutterstock; page 96 © Dani Vincek / Shutterstock; page 97 top © Worytko Pawel / Shutterstock; page 97 bottom left © Dja65 / Shutterstock; page 97 bottom right © Italianestro / Shutterstock; page 98 top © Scyther5 / Shutterstock; page 98 bottom © Adisa / Shutterstock; page 99 © Prill / Shutterstock; page 100 © Laborant / Shutterstock; page 101 top © Africa Studio / Shutterstock; page 101 middle © Africa Studio / Shutterstock; page 101 bottom © Miro Novak / Shutterstock; page 101 right © Iunewind / Shutterstock; page 102 © iBird / Shutterstock; page 103 © Deyan Georgiev Monticello / Shutterstock; page 104 © Monticello / Shutterstock; page 106 © Kunal Mehta / Shutterstock; page 107 top © Africa Studio / Shutterstock; page 107 bottom © Rob Wilson / Shutterstock; page 108 © Giuseppe Lancia / Shutterstock; page 110 top left © Thomas Barrat / Shutterstock; page 110 top middle © Winthrop Brookhouse / Shutterstock; page 110 top right © Tim Roberts Photography / Shutterstock; page 110 bottom left © Bullet74 / Shutterstock; page 110 bottom right © Monticello / Shutterstock; page 111 © Weldon Schloneger / Shutterstock; page 112 © Sigapo / Shutterstock; page 113 © VERSUSstudio / Shutterstock; page 115 © Somchaij / Shutterstock; page 117 © Iurii Kachkovskyi / Shutterstock; page 118 © Topseller / Shutterstock; page 121 © Maxim Ibragimov / Shutterstock; page 123 © Photographee.eu / Shutterstock; page 130 © Naypong / Shutterstock; page 136 © Olga Nikonova / Shutterstock.

Every effort has been made to credit the copyright holders of the images used in this book. We apologize for any unintentional omissions or errors and will insert the appropriate acknowledgment to any companies or individuals in subsequent editions of the work.

Acknowledgments

For sharing their time, knowledge and wisdom, I'd like to express my particular thanks to Arjen Hoekstra, Peter Gleick, Ruth Matthews, Stuart Orr, Jamie Bartram, Felix Dodds and Maude Barlow.

I am especially grateful for the years of love and support from my wife Renee and children Derek and Katrina.

Index

A

Africa
  access to drinkable water in, 46
  population growth, 37
agriculture, water footprint of, 113
Al-Kufrah (Libya) precipitation per year, 41
Alberta, virtual water exports from, 116
American cities, running out of water, 38
American Wind Energy Association, 15
animal feed, water footprint of, 19
Aoulef (Algeria) precipitation per year, 41
apple, water footprint of, 72
aquifers, major
  Great Artesian Basin, 30
  Guarani, 30
  Ogallala, 30, 31, 111, 114
  underlying major cities, 32
Aral Sea, 29
Asia, access to drinkable water in, 46
Association of California Water Agencies, 115
Atacama Desert (Chile), 41
Atlanta, water shortage in, 38
Australia, virtual water exports from, 116
Automobile Life Cycle Water Consumption, 98
avocado, water footprint of, 72

B

Bahamas, water-scarce nation ranking, 35
Bahrain, water-scarce nation ranking, 35
bananas, water footprint of, 75
Bangkok aquifer, 32
Bangladesh, 21
barley, water footprint of, 85
bath and shower, water footprint of, 57
beef, water footprint of, 76–78
beer, water footprint of, 91
Beijing aquifer, 32
biodiesel fuel, water footprint of, 13, 95
biofuels, impact on climate change, food
  scarcity, 14
bottled water, 18, 116
  increase in consumption of, 56

butter, water footprint of, 83

C

California, unsustainable water usage in,
  114
car washing, 125
cheeseburger, water footprint of, 81
Cherrapunji (India) precipitation per year,
  42
chicken, water footprint of, 77, 82
Chile, melting of glaciers in, 43
China, damage to environment of, 116
  freshwater extraction in, 39
chocolate, water footprint of, 89
circuit board, water footprint of, 61
Clarke, Arthur C., 22
climate change, 14, 36, 37, 114
clothes washers, energy efficient, 122
clothing, water footprint of, 20–21
coffee, water footprint of, 71
cola, water footprint of, 7, 90
Colorado River, water usage, 110
consumptive water use, by category, 105
cotton, water footprint of, 59
Czech Republic, water-stressed nation ranking,
  35

D

deepest lake in the world, 26
deforestation, impact on flooding, 37
Delaware, water use in, 55
Denmark, water-stressed nation ranking, 35
desalination, 23
diamond, water footprint of, 65
diapers, cloth and disposable, 49
diet
  meat-based, water footprint of, 78
  vegetarian, water footprint of, 79
dishwashers, 122
driest places on Earth, list of, 41
drinking water, guidelines for safe, 68–69
drought-resistant lawns, 124